Patrick Moore's
Practical Astronomy Series

Springer
London
Berlin
Heidelberg
New York
Hong Kong
Milan
Paris
Tokyo

Other titles in this series

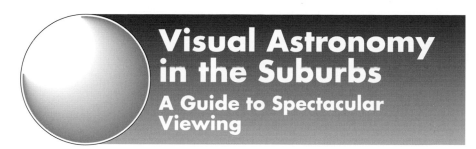

Visual Astronomy in the Suburbs
A Guide to Spectacular Viewing

Antony Cooke

With 143 Figures,
including 16 in color

Springer

British Library Cataloguing in Publication Data
Cooke, Antony
    Visual astronomy in the suburbs : a guide to spectacular
    viewing. – (Patrick Moore's practical astronomy series)
    1. Astronomy 2. Charge coupled devices 3. Image intensifiers
    4. Video astronomy
    I. Title
    522.6
ISBN 1852337079

Library of Congress Cataloging-in-Publication Data
Cooke, Antony, 1948–
    Visual astronomy in the suburbs: a guide to spectacular viewing /
    Antony Cooke.
        p. cm. – (Patrick Moore's practical astronomy series)
    Includes bibliographical references and index.
    ISBN 1-85233-707-9 (alk. paper)
    1. Astronomy–Amateurs' manuals. I. Title. II. Series.
QB63.C694 2003
520–dc21   2003042426

Patrick Moore's Practical Astronomy Series ISSN 1617–7185
ISBN 1–85233–707–9 Springer-Verlag London Berlin Heidelberg
a member of BertelsmannSpringer Science+Business Media GmbH
http://www.springer.co.uk

Typeset by EXPO Holdings, Malaysia
58/3830-543210 Printed on acid-free paper SPIN 10896631

Dedication

Dedicated to my dad, Nelson Cooke, great cellist and pedagogue, who nurtured my life and career, and only recently discovered for himself the wonders of amateur astronomy. As a small token of my esteem and affection, may this book inspire and guide him.

Special thanks to Jim Corley for helping me prepare the final draft of this book.

The author and his 18″ JMI Newtonian Reflector
on his viewing deck

Foreword

Suburban astronomy presents both a challenge and an opportunity to the amateur astronomer. The challenge lies in overcoming the obstacles of light pollution, haze and smog. The opportunity is in obtaining results that in the not-so-distant past would rival those of professional astronomers. This revolution in amateur astronomy is being led by the availability of CCD, video and image intensifier systems that perform at levels undreamed of by amateurs only a decade or two ago. In the case of CCD cameras, the equipment available actually outperforms that used by professional observatories less than a decade ago. Image intensifier systems offered to amateurs today, the Generation 3 Omni 4 product, is identical to the devices used regularly by professional astronomers. In fact, the Generation 3 system of today is actually superior in terms of signal-to-noise ratio, resolution, and photo response compared to professional systems used just a few years ago.

Image intensifiers for astronomical applications fill a unique role. No other product available to amateurs in the US provides the capability to view or image deep sky objects in true, real time (with no image accumulation time required). The increase in effective light gathering ability is improved on any given telescope by two to four plus magnitudes. There are, of course, "purists" in amateur astronomy, who refuse to use any electronic aid, be it a CCD video camera or image intensifier to assist in their observing. The rest of us can take a tip from the pros and use every device within our means to improve and enhance our ability to see deep sky objects. Tony's book should prove to be a valuable resource to those amateurs seeking guidance in their use of electronic aids, and especially suburban observing of deep sky objects with image intensifiers.

When configured for intensified video using frame averaging and a 7-inch refractor, an image intensified system such as the I Cubed is capable of resolving stars past 16th magnitude and galaxies to 15th magnitude in suburban (visual magnitude 4–4.5) skies. Amateurs should take note that the identical image intensifier system now available to them is used regularly on some of the world's largest telescopes at Mauna Kea and elsewhere.

Astrophotography is also greatly enhanced with image intensification. The photos I supplied Tony for this book were made using my 7-inch Astro-Physics refractor and a Canon D30 digital camera in conjunction with my I Cubed (I_3) system. With the ISO speed at 400, the exposure time was 6 seconds! While CCD cameras are capable of imaging to deeper magnitudes with a given telescope, one must not lose sight of the issue of exposure time. A skilled amateur, using a CCD camera, can usually obtain good results on only a handful of deep sky objects in a night of observing. A Generation 3 image intensifier, however, combined with a digital camera, can image a sky full of Messier and other deep sky objects in just a few hours. Many amateurs, using the I Cubed system, report that in average-to-poor suburban observing conditions, they are able to see many deep sky objects for the first time, compared to simple visual observing using the same telescope. The value of being able to point a telescope anywhere in the sky and instantly see to greater magnitudes cannot be overstated. Even some of the unassisted "visual only" observers find the utility of image intensified viewing to be a great assistance in locating "faint fuzzies", stars, asteroids or nebulae.

Allow me to share one of my own memorable observing experiences. About a year ago, I was observing in the region of M82 when, quite accidentally, I happened upon a very faint galaxy a few arc minutes across. I adjusted the contrast and brightness controls on the monitor and optimized the control box settings for the Astrovid CCD video camera. There, in the comfort of my home, I saw a tiny smudge of light whose photons had traveled for millions of years to arrive at the aperture of my telescope. I wondered for a few minutes about all the Solar Systems and perhaps civilizations that may have flourished within the borders of that small "island universe" and felt blessed to be observing at that fleeting moment in time. The experience seemed to me what astronomy, be it professional or amateur, is all about. I am fortunate indeed to be associated with the fine people who pursue amateur and professional astronomy, and particularly Tony

Cooke, who has taken the time and effort to present such valuable experience-based information to amateur astronomers who seek the insight that this book provides.

Bill Collins
April 2003

Contents

Chapter 1

Purpose and Fundamentals

The ability to observe the wonders of the universe is surely one of the greatest gifts from science to the layman. Much sophisticated equipment, once available only to the professional astronomer, has slowly entered the amateur's realm, and today the inclusion of advanced CCD technology has further revolutionized the possibilities. In seasoned and skilled hands, CCD equipment, along with an array of high quality telescopes, has made exquisite photographic images possible which equal or even surpass many great observatory photographs, even of the recent past. However, this leaves me thirsting all the more to see for myself, in real time, as many of those same sights as possible; there is still no substitute for the moment itself. With the exception of the Solar System, what we witness is not likely to equal, or even approach, the incredible wealth of detail, brightness and resolution that modern imaging technology has brought about. However, the eye's and mind's unique response to faint light sources still separates it from any other. All we need is to have visual access to these objects. In my own quest to have a connection with places viewed across time and space, the telescope is my spaceship, but its flights are potentially grounded within city limits by the challenges of observing from such an environment; many amateurs have given up and work only from remote dark sky sites. Indeed, for many, this has been a standard practice for years.

The primary aim of this book is to aid in the pursuit of exceptional live viewing, and from our own suburban environment. In the vicinity of the city, light and air pollution are our largest problems; urban humidity only adds to the lack of atmospheric transparency. From these sites, deep space viewing doesn't seem a remotely realistic proposition, let alone anything that could be described as spectacular. Suburban viewers are usually limited to glimpsing only the brightest Messier objects. We must face the dilemma: a lack of opportunity to pursue astronomy with a frequency equal to the passion. One remedy has been to use binoculars or richest-field telescopes to see some of the larger and brighter objects to greater satisfaction, but this is not the direction I wanted to pursue for this book. While binoculars provide a highly valid and enlightening approach, there is only so much access to deep space objects one can gain this way. I wanted to find whatever ways I could, with my telescope, to access at least some of what can be seen from remote dark sky sites, but from my home base. As it turned out, with the investment in some additional equipment, there is actually much you can see with your telescope in the suburbs. The results can be spectacular indeed! Of course, we should take every opportunity possible visit ideal locations, and look for the best conditions, but what about the rest of the time? Since my own text is designed as an attempt to reveal what can be achieved in far less than optimal (sometimes very much less than optimal!) circumstances, it is my hope that true suburban backyard astronomy may again become a reality for you. Therefore, if the occasional, or even frequent trip to an ideal dark sky site is enough for you, this book is probably not for you. However, I still invite you to read on and would hope you will enjoy its pages and ponder some other possibilities.

Of the two primary sky pollutants, light is the most troubling, and affects us mostly as a perpetual twilight that washes away the stars. This topic is rapidly becoming a *cause célèbre* around the world; you are probably already well aware of it, and many young people today have never experienced the awesome presence of the firmament under a truly dark sky. It is getting increasingly hard to escape the glow of some urban area, almost no matter where we live. Most of us simply do not have the time needed to drag our equipment to a better location whenever the bug bites us; for me, an occasional trip to the mountains just isn't enough to sustain the fire. As increasingly large segments of our populations become occupants of cities, or areas near cities, it is apparent that these hubs of civilization are becoming increasingly hostile places to look toward the stars, as they infect the night sky for countless miles around. In addition to light and atmospheric pollutions, the low elevations of most urban centers necessitate peering through the thickest and most turbulent layers of the atmosphere as well. Coupled with this, every city-dwelling observer has to contend with the frequent hazard of high buildings and obstacles, which restrict the viewable sky, leaving the horizon seldom visible. (I still do not have a remedy for this particular problem!)

For most people, the old-fashioned notion of observing the heavens the old-fashioned way, with one's own eyes, seems to be headed for extinction. However, it is possible to preserve much of this ideal through carefully applying the best that technology has to offer. I therefore approach observational astronomy with a somewhat different slant to the thrust of most textbooks today. I do not set out to show what can be done in great conditions and locations. Certainly you won't find in these pages a source of the usual sky charts, facts and figures common to so much available literature. I also don't try to espouse all the latest releases in astronomical software, which seem often to relegate actual observing to secondary importance. Finally, I certainly didn't set out to create yet another slick, gaudily colored, graphics-saturated volume of the type so common today; I often wonder if many publishers assume we have all lost our imaginations and intelligence!

Visual Astronomy in the Suburbs is intended to be used along with access to other reference sources, and at least a good celestial atlas. As a reference source, I strongly recommend *Burnham's Celestial Handbook*, Volumes 1–3 (see Bibliography) as your primary resource of detailed information. There are, of course, other materials available, including software resources, but in my opinion, *Burnham's* is still hard to beat. (How did just one man compile such an epic in his spare time?) Its remarkable wealth of information and infectious enthusiasm will complete or supplement the background that this volume does not set out to do. Although *Burnham's* is aimed at users of telescopes of up to 12 inches, the limitations imposed by our suburban environments make many of *Burnham's* listings of celestial sights impossible to see effectively (or even at all) when using substantially larger apertures than 12 inches. Because of this, you can easily be assured that you will not run out of sights to explore listed within its pages. Meanwhile, relegated to suburban viewing:

(i) What can be done to combat the difficulties of pursuing real time astronomy around our population centers, and to what extent?

(ii) Of countless sights in the sky, which of them can we realistically expect to provide spectacular viewing?

(iii) What measures can we take, what can we reasonably expect to see, and how will these sights actually look to us?

(iv) How can we best share and record our real time viewing, without the need for elaborate time-exposure imaging?

Almost all of the many fine books on practical astronomy for amateurs seem to make one assumption – ideal, or near ideal viewing conditions. Photographs and drawings by skilled individuals, in superlative situations, are presented to us. Often, the colors and detail represented by these images would no doubt lead the suburban novice to anticipate such views, live, through a newly purchased telescope. Now, the makers of many popular models of telescopes and accessories aren't in the business of educating the public about the challenges facing the would-be amateur astronomer. Also, the large boxes in which these scopes are packed (and even the ads touting them) often feature magnificent pictures of celestial sights, presenting an entirely false expectation to the uninitiated. Most of these unsuspecting souls must have a vision that they merely have to cart their shiny new scope outside, aim it, and see multitudes of heavenly wonders glowing brilliantly in the field of view. Such views aren't the case even at an ideal site! Is it any surprise, then, that for most people, a budding interest in astronomy soon wanes after the purchase of their first "department store" telescope? Add to this the simple fact that these cheap, but not inexpensive instruments aren't good for much in the first place.

More remarkable to me still are the sales of better-grade telescopes, equipped with on-board sophisticated controls, featuring thousands upon thousands of pre-set objects and even automated object location ("Go To" scopes) – whiz-bang gizmos to be sure. There is only one problem – their apertures typically aren't sufficient to show most of these wonders with any degree of satisfaction to their uninitiated new owners. This is even more certainly the case in the suburban locations in which they will likely be used. In most cases, the vast majority of these objects will be completely invisible! Most of their uninitiated users haven't even heard of "dark sky sites".

Some die-hards will tell you that visual astronomy from the city boils down to viewing only the Moon and planets. Since the Solar System is a major area of interest to me, and it provides a significant part of this writing, you can take it that I don't intend to minimize its significance for us in suburbia. It is also the one area of practical backyard astronomy that can be successfully carried out from the suburbs without any special provisions being taken. Admittedly, we may have fewer nights of steady air than a high altitude site, but there is no reason not to expect some first-class results from our suburban lairs. However, though there is much pleasure in the study of these subjects alone (some do it exclusively), most of us will ultimately want more, since we know it is there; it is only one part of the whole. Deep space, the true realm of the universe, beckons. To this end, if we live towards the edges of the city in more rural locations, things indeed do become more promising as the glow in the sky diminishes. Ultimately, however, some degree of the same limitations imposed on us by city life becomes a factor again in our consciousness; we know we are missing something, even though such a semi-rural viewing site might be a dream come true for the suburban or urban observer. Reluctantly, it must be acknowledged that even in these improved surroundings much potential for good viewing is lost, and for the same reasons. Therefore, I hope the direction of this writing will even be of interest to these somewhat better situated stargazers as well; they may actually have the greatest promise amongst us for radical improvements in their own viewing.

In order to be successful in your pursuit, I will assume that you possess an appropriate telescope for your surroundings. Nobody can tell you exactly what that should be, though the maximum aperture – of quality and practicality – should always be your strongest single goal. Dobsonian telescopes have certainly illustrated this concept wisely: bare bones, optical dimensions forefront. We have to recognize that different conditions and apertures at any given site will affect what can be seen and how well. (Where I live the bright light of the suburbs and frequent hazy conditions often will not permit some faint objects to show in my own 18-inch, which a good 6-inch in much better conditions will reveal. However, the 18-inch's resolution on any visible object naturally will still be superior.) I should stress that for deep space observing there is absolutely no basis to the enduring myth that a small aperture outperforms a large one in the type of sky conditions that this writing addresses. The larger aperture will always outperform the smaller. It is said that such apertures only amplify skyglow. This is true, but they also amplify the light source we are looking at! The large aperture can also handle significantly higher magnifications on these subjects without losing them in the background sky, and these higher powers will sometimes work to our advantage when detecting many faint objects, by throwing the background into deeper contrast.

So while I firmly encourage you to buy the largest telescope you can afford and use practically, be aware that a downside of the greater sizes will be their increasing negative reaction to unstable air, and the need for more time to cool down to nighttime temperatures. Fortunately, many of the problems we face from air currents and temperature changes only become serious issues at higher powers, and much of deep space viewing will be at its best with lower powers. It is also true that once the amateur observer has access to a moderately large telescope, say, 12 inches aperture and more, the differences with even sizeable professional observatory instruments are not nearly as drastic in unaided real time viewing as one might suppose, given equal viewing conditions.

Here are some general principles in regard to viewing conditions and our suburban environment:

- Haze and thin fog are not necessarily bad for lunar and planetary viewing. If these conditions result from stable and still air, the "seeing" may indeed be pretty good. The haze will sometimes reduce glare and act as a filter on these subjects. Viewing them is just as effectively realized from areas of great light pollution, and even bright nearby lights! All in all, they remain a suburban gold mine for us.

- These same conditions are ruinous for deep space, however. A good rule of thumb is to save your galaxy-hopping for the most transparent skies you are likely to have. They respond best to the clearest air; steadiness (the state of "seeing") is not nearly so important as it is for the planets and Moon. Total surface brightness for all deep space objects is a key; the suburbs render more difficult spread-out objects of high total magnitudes.

- Planetary and emission nebulae have an amazing tolerance to less than ideal air, and even though a perfect sky would be best, they can still put on a surprisingly good show in poor conditions.
- Reflection nebulae again need clear air to be at their best, as increasing haziness may obscure them completely.
- Star clusters will do better than most subjects in less than perfect sky, though naturally, transparency is always preferable. This is quite apparent in their ability to appear spectacular and to reveal any dark lanes.
- In my own location (Southern California), I have found that air currents moving off the sea onto the land are usually far more damaging to "seeing" than the reverse. Often a mild "Santa Ana" wind condition off the deserts brings the best "seeing" for lunar and planetary observing, as well as the clearest air. Viewing the Solar System at these times is sometimes breathtaking.

By my own definition, I chose to take the maximum "big gun" approach to the equipment dilemma, but bear in mind that my own location, astronomically speaking, is about as bad as it could be, despite the seeming endless blue skies of Southern California. It is well known that light and atmospheric pollution are already the norm in Southern California, which certainly makes deep sky observing anywhere near our population centers a major challenge. This has only become worse over the years. To compound matters, I live on a cliff-top at the ocean, with the resulting increase in humidity, haze, frequent night-time marine layer, and destabilizing ocean breezes; living here has necessitated some compromise for astronomical purposes. But since astronomy in most locations closer to the city hub (Los Angeles), would not necessarily be more effective, the only other alternative would have been to allow astronomy to rule all my life decisions and live out in the desert, far away from civilization! Nevertheless, through the addition of some of today's technologies (see Chapter 2), which give us a new lease on astronomical life in the suburbs, I am enjoying some wonderful viewing, in conditions that I am sure are substantially worse than most stargazers ever have to deal with. I am confident that equal, if not even superior, results can be obtained by much less means in less hostile skies. (It is true that over the ocean, light pollution does slowly drop toward the horizon – but near the horizon, haze is usually at a maximum!) I do, however, enjoy an uninterrupted view of most of the sky, and over the water, a totally unobscured view. This is indeed something of a bonus, particularly in viewing some of the Southern Hemisphere objects which are visible to us in the Northern Hemisphere.

Aside from our physical location, we also have to learn to see all that is present in the telescope image, if we are to be effective as astronomical observers. This is the subject of Chapter 3, and the need for it is true also even in the best of circumstances. It is all the more important in areas of viewing difficulties such as ours, where we are trying to extract the maximum results from the viewing situation. Looking through a telescope is not like looking at a picture on the wall, or television, much less the type of lavishly illustrated book on astronomy that we are all so accustomed to seeing. It is anything but the passive pointing of our telescopes and simply seeing what is there. The atmosphere is usually in a constant state of motion, more often instability. The light level of the objects we wish to observe, even under ideal circumstances, is low; the tiny scraps of light we are receiving is all that is reaching us from these extraordinarily remote places. City air, and especially light pollution, become critical if this is where we must observe most of the time. Since we are looking for details which are at the very limits of visibility, our eyes also need whatever degree of dark adaptation is possible, something made all the more difficult by nearby direct city lights. With patience and enough enthusiasm-driven commitment, we can acquire seeing skills; even then, we will still need some key technology.

Chapter 4 is concerned with techniques and methods we can use for recording what we see as it appears in real time. Hopefully it will furnish you with the means to expand the range of real time observing experience, and to share it with others. In addition to the chapters on the Moon (Chapter 5) and Planets (Chapter 6), I have devoted the largest portion of this writing to that almost lost denizen of the city dweller – deep space observing. Chapter 7 features a catalog and descriptions of the most successfully realized deep space objects for suburban dwellers known to me. Some of these may already be familiar to you as Messier and other objects, their inclusion probably not unexpected. More surprising to you may be those well-known objects (including many of Messier's) that do not make the list. Many of them are unsuccessful to view in the suburban environment; some have always been successful as binocular objects only, and this is not a book centered around binocular astronomy. Possibly most interesting will be those objects not so familiar to amateur observers. Here, also as in Chapters 5, 6, I have endeavored to provide very real visual impressions of what you may actually expect to see in typical suburban conditions for all those objects I describe.

In the approaches to observing in the standard literature and guidance available, it seems to me that disappointment is virtually guaranteed for any newcomer who is largely confined to the city. Aside from a suitable telescope, most of the objects in Chapters 7, 8 and 9 generally require the accessories alluded to already, along with which they may be successfully observed. The visual impressions and images included within the text are necessarily confined to those objects visible from my own latitude – in fact, a remarkable percentage of the total. Hopefully, in Chapter 7 these real time images will illustrate the essence of what you may expect to see, when viewing each of these objects. They attempt to convey something of the real time experience itself, and generally consist of live video images and/or drawings. (I should still emphasize, however, that there is simply no way to substitute a picture in any form, or a view on any monitor, equal to the live view through the telescope; this remains in its own realm.) Where it was not possible to include a video image (as opposed to a drawing), it was because the particular object does not respond favorably to such imaging, or is negatively impacted by my own circumstances. Maybe the relatively high minimum powers my telescope imposes made only a drawing possible, because of the much reduced field of view of the image intensifier (my lowest power regular eyepiece has a far wider field of view). Still in other situations, a drawing creates an entirely false impression. Star clusters are such items. For the most part, the eye is not better prepared by studying drawings of these objects. I have yet to see an illustration that actually looks anything like one, at least as far as conveying the visual appearance is concerned. Often complex structures are drawn that seem to have no resemblance to anything I have ever seen, while missing features that I have seen or recorded by video camera. However, for most deep space objects, I have been able to include both forms of image. I hope the reader will gain the most complete expectation possible from the examples. This is obviously a moving target, depending on a combination of your own circumstances and equipment, but I believe the illustrations I have selected for each object will more properly prepare you for what you are likely to see from the suburbs than most other materials you can find. A large proportion of the objects totaling the listings of all three chapters will be visible from both Northern and Southern Hemispheres, and at many latitudes.

The sights of Chapter 7 have also been carefully chosen to give the backyard astronomer a catalog of objects throughout the year likely to delight visitors. How often are these often unfortunate people (our victims?!) left wondering what we get so excited about, while we fumble about trying to find a successful view of something – anything! Either that, or we resort to the same old tried and tested meager handful of "sure things", in the hope we can convince our visitor(s) that "they're all like that!"

The full extent of any deep space object will likely be seen only in time-exposed images. Some mental adjustments will have to be made in order to know what you are seeing when it comes to viewing them live. You should study as many observatory and CCD images of these objects as possible in advance, so by the time you compare what a quality exposure of a given subject reveals, you will make out more from the real time view, and the experience becomes all the more satisfying. Tiny pieces of faint detail assume great significance when you know what they represent in their veiled suggestion. Where an image I provide seems ambiguous (in showing whatever I could witness or capture), I also include a graphic with the object's fuller extensions indicated, so it will make more sense.

Common sense also tells us it is not only improvements in optical technology since Lord Rosse's time which make it no longer necessary to have a 72-inch aperture to see spiral structure in galaxies; more significantly, it is also because we know the spiral structure is there. But here is a critical caveat: we must strive to keep our eyes "honest" and not to let their "education" fool us into believing we are seeing something which we actually are not. (The "canals" of Mars are amongst the best known examples of this.) Conversely, drawings of planets made before the era of space exploration are remarkably different in character to drawings made today. This serves as one of the best lessons for us in "eye education". In addition, it is routine now to see details, and above all, to have overall impressions that were often never even mentioned in the past; through the telescope our impressions can be forever changed once we know what to look for. Since the advent of spacecraft and CCD imaging, I also don't believe that it is only due to three decades of relative calm on Mars and Jupiter that there seems to be a remarkable degree of "stability" on the surfaces of both of these worlds! I believe that at least some of this may again be due to our eyes becoming more accustomed to knowing what is really there.

Chapter 8 features my own carefully screened secondary catalog of other deep space objects definitely worth taking the trouble to look for, along with certain relevant information, and illustrations of many of the best examples. They are within reach from many suburban locations. Depending on your own location and viewing conditions, they may even become part of your own personal best list. For me, my own viewing circumstances do not quite allow them inclusion in the "best sights" catalog of Chapter 7, and my primary interest lies with objects that reveal themselves as more than just a blip or smudge in the field of view. I will admit that a few were a close call, but for better or worse, they were relegated to Chapter 8. Both listings (Chapters 7 and 8) represent objects, compiled from my location, which should be possible to view in suburban conditions, the first list (in Chapter 7) being the most likely to be in the realm of the spectacular. And of course, many of these same objects are visible to observers in the Southern Hemisphere. I have excluded whatever I feel will more than likely lead to disappointment and frustration rather than success. The listing in Chapter 9 completes the survey for viewers living in the Southern Hemisphere, and features a supplementary catalog of outstanding objects of the South, from the same standpoint of suburban viewing. It was not possible to include comparable depth of information on these objects, or images of them, but by the limited size of this final catalog observers in both hemispheres will realize just how much of the total sky can be seen from each.

The objects listed in Chapters 7, 8 and 9 have been arranged in the order that they progress across the celestial sphere; these catalogs are not so large as to make it unwieldy. To any observer using this book as a reference, this is designed to be more useful and practical than arranging them in alphabetical sequence. Alphabetical listing would require constant thumbing back and forth through the pages, when objects relatively near to each other in the sky could be accessed at the same time more easily by grouping them according to right ascension. I believe practical observers will find this system preferable in this particular instance.

Chapter 10 raises a few final comments and thoughts. The appendices list some of the more prominent manufacturers and suppliers of equipment relevant to this writing, along with some approximate cost guides, technical information, and other commentary.

Regarding many of the better-known "binocular" items, many do indeed appear on the second list, especially open star clusters. Through a telescope's much more limited field of view, these are often too spread out or faint to be impressive in the sense that Chapter 7 was compiled. This factor is compounded by increasing aperture (and hence typically shrinking field of view), though depending on your own circumstances and telescope, you may wish to add some of them to your personal "best" list if they turn out to warrant inclusion. In my opinion, though, most of the globular clusters simply outclass the open clusters anyway, when it comes to seeking out the absolute best that the sky has to offer, and unsurprisingly, some of the finest examples of these feature quite prominently in Chapter 7.

I am leaving alone the subjects of double and variable stars. They certainly represent a major part of astronomical interest and study for many people, but in my view, they do not represent the spirit or intention of this writing. However, I am sure some viewers might include them in the realm of the spectacular; maybe they are. You must determine this for yourself, but please forgive any apparent lack of acknowledgement that I may imply. At least many doubles and variables present few observing problems to the suburban dweller, and no special means to combat the problems common to other deep space objects may be necessary for successful observing programs. As major areas of study, doubles and variables are best left to the literature that covers the field, and would not be served appropriately by the approaches discussed here. If your interests extend to meteors, asteroids, occultations, comets and novae, they are all certainly within the realm of suburban viewing, especially when incorporating what this book outlines. Nevertheless they are outside the scope or intention of this writing; at times however, comets and novae are indeed capable of providing us sights that could be described as nothing less than spectacular, and as impressive as anything else contained here. At such times, you may wish to experiment with the same approaches discussed here, and your results may indeed be memorable. However, unless you are a hunter of these phenomenon, you will need to access monthly magazines and Internet sites, such as International Supernovae Network, if you want to find them without the enormous research and investment of time they will otherwise demand.

I am also leaving the observation of the Sun untouched. I do not venture towards it because I have not made a direct study of it, partially out of fear of the consequences of error or equipment failure! Even using one of the excellent solar filters available today, I am all too aware that this is the only thing between my eyes and the staggering solar blast that would be reflected off 18 inches of aperture in the event of failure. I simply cannot overcome my natural distrust to any such filter, wedge or the like, in spite of all the evidence of their safe usage. Projected solar images would be out of the question, with the furnace that the large aluminized surface of my scope would generate. A normal practice is to substitute an unaluminized primary for the purpose, but this naturally limits the objective's use to solar observing only, and it is not a practical option in larger sizes to have two objectives. In any event, since solar observing is a whole field of study by itself, I will leave it to others to guide you if this is a field that is of interest. Certainly its application is undiminished in the suburbs, but do not undertake it lightly; your equipment and expertise cannot fail you.

Chapter 2
Practical Applications and Viewing Aids

My own involvement with telescopes and astronomy goes back to childhood, starting with my first telescope, a tiny refractor. Working my way up through somewhat larger sizes, I eventually saved all that I could put together, to buy my first real astronomical telescope: a top-of-the-line Charles Frank 4-inch Newtonian reflector. I think I was 15 years old at the time. In the 1960s, this company was one of Great Britain's most respected manufacturers, and has apparently long since faded from view, like so many others from that time. I still recall the massive and workmanlike German equatorial mount, something other manufacturers could have learned from, even to this day. (This was no generic mounting of the type that routinely graces multiple manufacturers' products across the globe today, good though these may be.) For astronomy, this was the first true defining moment for me, and it seemed like having a piece of Mount Palomar, a true claim to be a real amateur astronomer. Of course, 4 inches of aperture in a reflector will only show you so much, and it wasn't too much later that the beginning of aperture fever hit me, along with a fascination for building something for myself, born at least partially out of financial necessity. With a strict budget to work within, I set about the task with expert guidance and inspiration from *Amateur Telescope Making*, Volumes I and II, edited by Albert Ingalls, and successfully built an $8\frac{1}{4}$-inch F8 Newtonian reflector.

This period of home-grown telescope construction turned out to be one of the most memorable periods of my life, and I regard it today as pivotal in the lasting passion for the hobby of astronomy that has held on to me ever since. It is truly regrettable that the abundant supply of inexpensive, and sometimes quite decent instruments flooding the market today has, by default, deprived a great many people of this kind of wonder (see Appendix A). Anyone who is the least bit bitten by the telescope bug owes it to himself or herself to read *Russell W. Porter*, by Berton C. Willard, 1976 (Sky Publishing). This volume, better than anything else, encapsulates the degree that all TNs (Telescope Nuts) are in debt to this one man. Our entire hobby would likely never have come into popular being; the cornerstone contributions to the *Amateur Telescope Making* volumes would not have been written, nor would the books even have existed, without the single-handed visions of this one man. And this is only part of what he did.

By the time I had become a telescope builder, I had already grasped the concept of massive, thanks to Porter, and most of my 8-inch telescope was created from unlikely components. However, it did have some commercial parts, like a Charles Frank helical focuser and secondary spider; some other parts of the mounting I bought used. At the time this seemed like a truly sizeable instrument. In a way it was, since light pollution in suburban London, where I spent some 15 early years, was far less of an issue then than in the situation I find myself today. I remember some very good nights from time to time (yes, even in England!), when that trusty $8\frac{1}{4}$-inch gave a very good account of itself. This was the telescope that "qualified" me in my mind to participate in main-stream amateur astronomy. I was there! I don't think, however, I'd be very happy now with it in my present situation, and none of my telescopes up to this time featured powered tracking.

Some years later, when I had relocated to Florida (where I lived for some six years), I built a much more ambitious scope. This time it was to be a $12\frac{1}{2}$-inch F9 Newtonian, which I had designed with viewing the planets primarily in mind. It was an open frame affair with a huge equatorial mounting, and I took great pride in building every part of this myself, the only exception being the eyepieces. I will never forget my first view of the Moon through it. I could hardly believe my eyes; the view was stunning, and it turned out (probably through some degree of luck) that I had fashioned a highly accurate optical system. The very long and unwieldy telescope did provide me with some spectacular views of deep sky objects as well, and my location in the suburbs of Tampa was probably no worse than many places today. This is not to say that I was unaware of increasing difficulties caused by skyglow. Even then, in a city far smaller than London, there was more of a light problem to contend with, not to mention much greater humidity and haze. However, at the time I never did acquire any light filters or other means to help combat the problem; these were only newly entering the marketplace at the time. Additionally, the telescope was securely attached to the ground in four feet of concrete. This made the occasional trip to darker sky locations out of the question, even if I could have transported the scope itself.

The following four sections detail steps I have taken which I regard as central to the entire pursuit of real time astronomy in the suburbs. You will likely be able to achieve much the same results with lesser means, depending on your specific circumstances, but I am certain some form of each is key.

Big Gun Number 1: Aperture and Telescope Choices

Upon briefly relocating to Chicago, my $12\frac{1}{2}$-inch lay dormant in the basement, and it was not until finally relocating to the Los Angeles area in 1984 that my old interest was slowly reawakened. I wanted aperture, and since aperture fever is a common malady amongst amateurs, a "mere" $12\frac{1}{2}$ inches was no longer enough! Coupled with this, the light-polluted sky in the area seemed to demand a new approach. After some time (actually, years!) of wrestling with the concept of buying a commercially made telescope (for me, something seemingly out of step with the hobby after all this time; see Appendix D), I finally reconciled myself to the dilemma and made the plunge. My situation in California resulted in the purchase of an equatorially mounted 18-inch F4.5 Newtonian (JMI NGT-18). Owing to its unique design and emphasis on portability, it can be moved around by one person, me (!), with the greatest of ease. I knew I could never have built anything this good, refined, or practical, and that I had made the right decision.

The NGT-18 is available with wheels and handlebars, and with these attached, it is not unlike moving around a loaded wheelbarrow. This portability was important to me, since I wanted to be able to move the telescope around my viewing deck, and periodically, of course, take it to dark sky sites. For ventures away from home, the scope breaks down in minutes to a size that will fit in many standard vehicles. This relatively compact behemoth was my first big gun to counter light pollution, aside from the fact that I wanted the larger aperture for all the reasons that amateur astronomers tend to want ever larger apertures! For a practical limit, this size, and more particularly the specific telescope variety I have, is about the largest equatorially mounted telescope one person could readily move alone.

Regardless of aperture, it still remains essential that we try to shield our scope from any local lights, as their resulting glare will destroy any other good attributes that may remain at your site. Even where no lights are in view, it is also important in these surroundings that open-frame telescope tubes are covered with light shrouds to block stray light, one of the biggest robbers of contrast. In suburbia, where the sky is never really dark, you also never know where stray light is being reflected from, no matter how slight it may seem. Remember, we are gathering (according to the aperture of the telescope) all light that enters the telescope and reaches the eyepiece.

It is also a very good idea to baffle effectively even closed Newtonian tubes at the eyepiece end, opposite the eyepiece; also behind the primary. Do this with the darkest, non-reflective material you can find. It is even more important to be sure that the upper end of the tube extends far enough to block out stray light at the eyepiece end as well. Simple extension baffles, again opposite the eyepiece, made of the simplest of dull black materials can remedy this problem. If in doubt, conduct this simple test: with no eyepiece in place, look into the telescope with your eye close in to the focuser, carefully looking all around. If you can see anything peripherally outside the telescope tube, even the sky, you will instantly know stray light can enter the field lens of the eyepiece. Make sure the extending baffle blocks this peripheral view, as most focusers are placed too near the top of the tube. Behind the primary, make sure that any stray light is not able to enter the tube from here as well, but be careful not to block the airflow.

The best advice: buy or build the largest aperture high quality telescope you can afford, equatorial or not. It is far better to forgo all the "bells and whistles" for light gathering power. Include as many of these "bells and whistles" as are useful to you, but only after aperture, as this should remain your priority. But also bear in mind that larger sizes will deliver high magnifications effectively on fewer nights, and are only of value if readily usable, practical, and suited to your own situation. Remember that, inch for inch, quality made Newtonian reflectors buy you much more than other varieties, and deliver outstanding performance. Personally, I tend to favor that design over most others anyway, and it has the wonderful feature of its eyepiece being located near the top of the tube and to the side, about the most accessible place it could be. I have always been amazed how this feature has been denigrated by some (perhaps by those popularizing other designs?). Except in cases of great focal length and very large apertures, the Newtonian generally provides the maximum comfort when viewing. I should stress, though, with equatorially mounted Newtonians some form of eyepiece repositioning is essential, so at the very least be sure the telescope has a tube that can rotate on its mount in order that the eyepiece can be positioned comfortably, depending on the object's location in the sky. Without this feature, for much of the time there is hardly a Newtonian that will not seem like a contortionist's best friend. With some designs, including my own JMI reflector, independent rotation of the upper nose assembly places the eyepiece exactly where is most desirable, a delightful arrangement. The collimation remains remarkably true through the entire rotation. Even with 18 inches of aperture at F4.5, the low-slung split-ring mounting (we have Russell Porter to thank again for this design; I really do have "a piece of Mount Palomar" now!) enables what could be an otherwise unwieldy tube length to necessitate, for me at 6 feet 1 inch, only one step up on a kitchen stool when pointing at the zenith. Another advantage of the split-ring design is the large diameter turning circle of the polar axis (36 inches). For tracking, this results in a very smooth and virtually periodic error-free motion. It also enables larger telescopes to have their centers of gravity placed so low that construction can be less massive than would be expected, and makes portability (or at least, one-man relocation) possible.

Many compound telescopes, better termed catadioptrics, (Schmidt–Cassegrains, Maksutov–Cassegrains, and now even Schmidt–Newtonians) may be convenient and readily portable, but in my view, and especially in the case of Schmidt–Cassegrains, they frequently combine added expense with many of the weaknesses we are trying to avoid. The chief advantage they bring is compactness, which has great merits if we consider the fact that most deep sky observers want to access sites far from home. They also tend to be offered in packages that include many sophisticated features, partly made possible because of their compact size. But it cannot be said that all of this convenience comes at no expense to image quality; this includes decreased contrast, due to a large secondary mirror, and an increased challenge in obtaining perfect focus due to a far greater sensitivity to slight movements of the focuser. Other drawbacks, not often discussed, are greater needs for stable air (owing to diffraction effects also caused by the large secondary), and problems of potential image shift when focusing (caused by the primary mirror moving for focusing instead of the eyepiece). This is not to say the best-engineered compound telescopes are necessarily bad, but for sheer performance, most designs remain somewhat inferior to simpler optical configurations, even when we are talking about the finest examples available.

However, there are greatly superior compound telescope designs, including the Maksutov–Newtonian. This hybrid configuration is outstanding in every way, but substantially more expensive than most telescope types, and often difficult to find in larger sizes. But Mak–Newts do offer stunning performance that is hard to beat, and typically, inch for inch, likely will be superior even to our old friend, the Newtonian. Not so the Schmidt–Newtonian, which still suffers from an overly large secondary mirror. There are still other optical configurations and designs, such as the classical Cassegrain or the sophisticated Ritchey–Chretien, that have their own strengths and weaknesses, but bear in mind that they come at much higher expense than most people are able or willing to pay, and may not be suited to our viewing needs. They are likely to have to be custom built, since they are not generally available off-the-shelf. For most amateurs, it is probably best to stick with the more widely used designs.

Apochromatic refractors of the highest grade have many advantages all of their own: color-free images, superb contrast owing to the lack of diffraction effects from a secondary mirror, maintenance-free use, and absence of the Newtonian's coma and convection currents. The best examples are wonderful instruments to be sure, but the same money will buy you a top-of-the-line reflector usually several times the aperture. All refractors require you to look through the eyepiece from the bottom, which always causes problems at the zenith, and require tall mountings and star diagonals. Their likely length and solid tubes guarantee weight.

JMI has just produced the prototype of a new instrument, which combines the attributes of Newtonian telescopes and the ease of binocular viewing. This is a 6 inch, F5 Newtonian binocular telescope. Because both eyes are receiving the image and are able to view it in a relaxed fashion, this type of instrument tends to produce a gain in real time magnitude perception. Planetary detail is easier to determine, and a strong sense of three dimensions is miraculously achieved, even though there is no way that the telescope truly can be producing it. The Moon will appear to be hanging in space, with its cratered face jumping out towards you. Even deep space objects will benefit in similar ways, with globular clusters providing spectacular results. For your purposes, it just may fit the bill in many suburban and other surroundings, but remember, every viewing accessory needs a duplicate (for each eye), and financially, some of those I detail in this writing become unrealistic. From JMI, it is bound to be sophisticated and beautifully made, and as such, do not expect an economy scope!

With short focus reflectors two additional things become especially critical. This is also true with catadioptric systems (something often overlooked when these telescopes' performances are wrongly criticized). Unsatisfactory results will occur if their optics are not in a really accurate state of collimation; I cannot emphasize this enough, and although it might seem like an inconvenience, it is really very easy to take care of routinely once you have mastered the system, (also after the correct positioning of the secondary mirror has been established and fixed). In the performing of collimation for telescopes of short focal ratios, the tiniest adjustments will prove to make "night-and-day" differences, so do not make more than fractional turns of the primary mirror's collimating screws. This is one of the most important points I have to emphasize, and adjustments of the primary mirror in larger increments than this are one of the reasons so many people experience difficulties with collimation. For a Newtonian, the use of a Cheshire eyepiece and sighting tube will make this procedure more straightforward, and is actually a necessity with short focal lengths. Laser collimating devices available today may produce fractionally better results with careful use, but they are by no means essential, and may actually be less satisfactory to use than the Cheshire. (The better Newtonians today retain their collimation well, even after moving the telescope frequently.) For touch-up adjustments, limiting this to only two of the three screws will prove to be a great time saver. This is more helpful than it sounds, since it reduces the opportunities for error and guesswork. Catadioptric systems may require professional collimation if it is needed, although with careful treatment, they seldom go out of adjustment.

On the topic of short focal length telescopes, the other necessity is to buy highly specialized and corrected (also expensive!) eyepieces, such as are offered by TeleVue. These are rather massive multi-lens units, designed for today's fast focal ratio telescopes and catadioptric systems. Before they were available, the short focus Newtonian and others were largely regarded as near hopeless basket-cases, only suitable for low magnifications and wide field views. Also be sure to consider the size of the exit pupil of the eyepiece to be used, which is a factor of magnification and aperture. This results in a minimum power for any given telescope. In order to take advantage of the full telescope aperture, anything greater than a 7 mm exit pupil is more than most dark adapted human iris' can open. Calculate the exit pupil by dividing the aperture of your telescope by the magnification used. An eyepiece which is theoretically unsuited to a particular instrument may still produce highly pleasing views, without the full aperture being utilized. This is because the gain in visual brightness may be less than the actual loss of effective aperture. Usually, reflectors and catadioptrics have a much lower tolerance for eyepieces yielding too large an exit pupil, and should not be more than 7 mm. This is because of the secondary mirror obstruction, so be careful. You might end up with a fine view of the obstruction's silhouette!

Big Gun Number 2: Light Pollution Filters

The second big gun in the war against light pollution was the purchase of light filters. These have turned out to be a nearly indispensable aid on many subjects, as they will often throw the subject into stark contrast against the sky. I have found that narrowband filters (as opposed to broadband or other) are preferable for most suburban purposes, as they allow the transmission of light from the target object and suppress other wavelengths, including sodium and mercury vapor streetlamps. In my own situation, I get excellent results on the majority of deep space objects with Orion's 2-inch Ultrablock filter (Orion of USA, not the venerable Orion Optics of England!), in combination with a 27 mm TeleVue Panoptic eyepiece, for a magnification of 76×. While I tend to characterize the mainstay and style of the Orion catalog as being primarily aimed toward novice and intermediate "stargazers", the company nevertheless does include some fine products for a more advanced clientele. Broadband filters are also available, and likely to be less expensive. They tend to make the view more agreeable, but generally do not reveal much more than was already apparent before. As with all things, trial and error is always the best way to find out what works best, on what, and for you.

Light filters are not for everything, although on some things they make all the difference between invisibility (or just a ghost of an image), and a strikingly contrasted view, even from dark sky sites. Also, they sometimes help in the search for certain faint deep space objects, as they darken the background sky (although many background stars will disappear from view as well). You might also want to investigate some of the other leading brands of light filter, and namely the much-respected Lumicon brand. I would recommend their UHC filter as very similar in performance to Orion's Ultrablock. Lumicon also markets some interesting variations of the standard narrowband filter, with very specialized usages in mind. (In the confines of suburbia, though, you will need to experiment with these to see if they provide any benefit; this may be marginal, and on only a few subjects.) There are other makes of light pollution filters, of course, with individual preference determining which one is best for you, and also available now, light filters for image intensifiers (the subject of the next "big gun"). More in the next section.

Big Gun Number 3: Image Intensifiers

The third and most important weapon in my city viewing arsenal, after the telescope itself, was the purchase of a modern third generation image intensifier, especially made for astronomy (the remarkable I_3 Piece by Collins Electro Optics). The beauty of the Collins device is that you use it just as you would a conventional eyepiece. Image intensifiers require no image processing, monitors or separate controls.

Image intensifiers *per se* are hardly new to astronomy, and have been written about quite regularly since the late 1970s; actual applications in real time observing have been only marginally discussed in publications for amateur observers. New generations make them more valid than ever, and they are far from being fully recognized for what they can offer. You will soon realize that they represent a significant focus of this book. While expensive to be sure, I must stress that image intensifiers are the single most significant accessory I know of for our dubious surroundings. They will cost you less than most good CCD equipment with all its trappings.

The CCD revolution that has taken place has resulted in the intensifier taking something of a back seat to the pursuit of acquiring of highly resolved astronomical images. Nevertheless, the central theme of this book remains focused on real time observing. I still chuckle at an advertisement which actually boasted of the beautiful images that could be recorded by a CCD camera attached to the owner's telescope while he spent time inside watching TV with his family! Another proclaimed the wonders of recording near-limitless numbers of deep space objects, while our supposed astronomer was able to sleep peacefully through the night! Outside of comet, asteroid and supernova hunting, doesn't this approach miss the whole point of amateur astronomy in the first place?

Even though it is possible to download amazing CCD images off the Internet, none of this produces the joy of seeing these objects with one's own eyes. While image intensifiers do not work on everything, their effects are so significant on so many subjects that they bring much of what we have lost in the suburbs back into visibility. Where they do not work, it is typically because an object emits its predominant spectrum around a wavelength less favorable to the intensifier, negating its value. This might even result in negative results, since skyglow will also be increased relative to the object. Disappointment likely will result with their use on most reflection nebulae, objects that usually reflect light from the blue end of the spectrum, the portion in which image intensifiers are least responsive.

One of the primary differences between an image intensifier's range and that of the eye is that the intensifier responds to a wider spectrum, and is strongly biased towards red and infrared, including the celebrated Oxygen III emission line of 496.32 nm. When the intensifier is able to gather light concentrations nearest to its highest response, results are at their most startling, and can reveal an object in a dramatic way. The most advanced intensifiers available to the public are designed to minimize "noise", and thus are better suited to enhance contrast between object and sky, particularly in conditions that might not have worked so well with earlier designs. One should still be prepared for some slight noise and "snow" in the image, and though these devices achieve decent focus, star points become noticeably larger disks with increasing magnitudes.

Because of known information on the makeup of the light coming from any given object, one should theoretically be able to predict how effective the intensification will be. In reality, part of the fascination is that one never knows exactly how well it will work in practice, and it is quite possible to be completely bowled over when least expecting it! I will never forget my first view of M82 with intensification. With such a large and bright galaxy of its particular spectral makeup, I knew it would be good. In practice, though, I was stunned by all the tangled detail and brightness suddenly visible to me, where only a formless smudge had been before. The range of light, shading and dark veins was truly astounding, regardless of any expectation I had.

The image seen through these devices is also right side up, as in terrestrial telescopes. Seen as an image on a phosphor screen, it is monochrome green (Figure 2.1). At first this might seem to be a distraction, but it is amazing how quickly the eye gets used to it, and effectively interprets what it sees as black and white; even a partially dark adapted eye reduces awareness of color still further. Green pictures would be a distraction (although striking!), because in the context of the page, and in normal lighting conditions, they will only be seen as green! For this reason, all intensified real time images are reproduced here black-and-white. (Those by W.J. Collins, digital camera short time exposures, are in color, to convey the initial live impression through the telescope.) The color of the intensifier image is not just a quirk that we have to live with; it is selected as the best frequency for the eye in dark conditions. The image, of course, includes all frequencies, even those not normally visible, or marginally so. This can give us a significant boost in sensitivity to regions of the spectrum that have not been gobbled up by pollution and skyglow.

All of the above might lead you to believe that the view through such devices would seem completely artificial, and far detached from the spirit of real time viewing. Surprisingly, this is not the case, and intensified viewing seems as natural and real as any other, once you get used to the immediate differences. Only a limited amount of dark adaptation for the eye is needed, a nice by-product, although this can be some-

Figure 2.1. Example of real time image. M16 in Serpens, intensified image. Courtesy W.J. Collins.

thing of a problem when alternating regular and intensified viewing. Despite what you may have read, you should know also that averted viewing of many objects with image intensifiers will indeed reap considerable rewards. This is never more the case than when looking for "dark lanes", and other aspects of contrast. I don't believe the eye changes its sensitivity to these things just because we are using an image intensifier! The practical effect in light gathering (not resolution) is frequently comparable to doubling or even tripling your telescope's effective aperture, and remember, aperture is the thing! Imagine being able to look directly into the eyepiece and seeing many well-known objects looking so recognizably close to their well-known portraits! Additionally, as mentioned before, there are some recent advances in light filters especially for image intensifiers, which are designed to increase their effect still further. (See Chapter 3.)

You will also probably hear the argument that it is just a matter of time before technology will outdate any intensifier equipment you buy and its value will drop like a stone; an important issue considering the costs involved. I don't believe this is an accurate assessment. This sophisticated technology is not like the latest thing in consumer electronics or automobiles. It is not like a new TV or DVD player. Actually, there is already a fourth generation intensifier tube in existence, though its slight benefits for astronomical use would probably not be worth the much greater additional cost to most people. You will see from any research you undertake that earlier generations of intensifiers are still very much in existence, and even now, good quality examples for our purposes are not exactly cheap. So it is fair to say that the best technology available today will likely still be giving good results years from now with just a little care, and that further developments will not automatically reduce today's equipment to obsolescence.

Collins Electro Optics presently has no international distributors, so check with them for specifics in your own situation, as certain export restrictions exist at the time of this writing. However, there are other highly viable options available, though none so ideal as the Collins. Even a used unit might fit the bill. I am not sure if it would be possible at this date to find a Collins I_3 on the secondhand market, but you may be able to find another make from an earlier time. In the early 1980s, a company out of New Jersey, by the name of Electrophysics Corporation, offered complete eyepiece/

intensifiers much like the I_3 in concept, only far bulkier. Another company, by the name of Varo, popularly marketed intensifiers which were commercially adapted for use with telescopes under the trade name Noctron. It may still be possible to locate one of these older units, and even a later and upgraded model, in the used marketplace. Perhaps one might provide worthwhile results. (I do stress that these older makes are earlier generations, and you should not expect equivalent performance of the I_3. The signal-to-noise ratio of the I_3's Generation III tube is many times superior to what preceded it, not to mention its improved spectral response. This becomes ever more important when searching out faint objects, and particularly in our worsening suburban sky conditions.)

Meanwhile, Electrophysics Corporation is still alive and well (see Appendix A), offering a wide range of intensifier products (Generations II through IV) that can probably be adapted to astronomical use with a little ingenuity. Noctron products (with second and third generation tubes) are now marketed by Aspect Technology and Equipment, inc. in Texas, but are manufactured as complete night vision scopes of very low magnifications. Some of these products also may likely lend themselves well to adaptation to astronomical use. There are many other suppliers of image intensifiers in Europe as well, some of them offering Generation III intensifiers. So there is a wide choice available. Bear in mind that the more advanced the generation, the higher the cost is likely to be. I do not know which of these companies would be willing to work with a prospective purchaser of components only, though I believe most will. Regrettably, to my knowledge, no other intensifier product today is ready to go as a real time astronomical device. Apparently, only Collins provides such a thing today, in 2003, if you should be so fortunate to acquire one. It is a beautiful unit indeed, featuring a ready-to-go combination of premium intensifier technology, specialized integral eyepiece component, and $1\frac{1}{4}$-inch or 2-inch adapter. There is also a line of high grade accessories. Apparently only one company at the present time is willing to take the sizeable business risk of presenting such equipment, long available to professional astronomers, to the amateur. Such risk taking requires a considerable vision of what possibilities this technology can offer. Bill Collins has carved out for himself a unique place amongst those who have helped propel amateur astronomy to new horizons; he deserves recognition as today's lone pioneer, and has certainly had no help in many seemingly likely places. Philip S. Harringtons' *Star Ware* (published by John Wiley), touted as the ultimate guide to all things for the amateur astronomer, doesn't even mention image intensifiers. Collins' products have even met with outright hostility in some amateur astronomers' circles. (For some manufactures and distributors in the USA and Europe, technical information on the Collins system, and third generation image intensifiers, see Appendix.)

Intensifiers project a tiny image onto an integral concave "screen", which needs to be magnified and rendered flat. The eyepiece component made by TeleVue for the Collins I_3 is designed to produce a flat image and magnify it to the specifications, something like a sophisticated magnifying glass. In the case of the I_3, the units are supplied producing either the eyepiece equivalent of 25 mm or 15 mm focal lengths. The two versions are used like an eyepiece of similar focal lengths. The unit can readily be attached to a Barlow lens for different magnifications; more on that later. Although their fields of view are substantially narrower than today's new breeds of multiple lens eyepieces, they are certainly more than adequate for our purposes. In adapting intensifiers other than the I_3, it ought to be possible to select an eye lens to produce a usable image across at least a portion of the field, but experimentation is the keyword. The only other issue is providing the unit with a $1\frac{1}{4}$-inch or 2-inch tube adapter to allow it to fit your focuser like any other eyepiece.

As with all astronomy, stray light is also an issue for image intensifiers. Bad as any reflected light will be on contrast in normal circumstances, when this light falls into an intensifier, the results are disastrous. All such light will be amplified tens of thousands of times, and what may seem insignificant to general viewing becomes like the light of day in an intensifier. Such events can also damage the unit itself if they are bright enough. Be sure to baffle all potential stray light sources from entering the optical path, and above all, out of the image intensifier.

If you decide to pursue image intensified astronomy, then at the very least buy the best intensifier you can afford. I feel strongly that you will find that in the suburban sky conditions we face, this tool will become second only to the telescope itself. Still more innovations are available to help crack the problems we face, but none of them to date (except filters – on certain subjects) can enable the user to look directly at so many deep space objects, and experience them so effectively through the telescope, live, and in real time. Although clearly at a disadvantage, the reader-observer without access to such equipment, or even a good light filter, need not despair to the point of thinking all is lost. If he has easy access to a location significantly better than mine, he can take much of the information in these pages, and apply it to unaided real time viewing, with significantly more modest equipment. He can still expect a decent show. (See also Appendix A for more information and specifics.)

Big Gun Number 4: Video

A very good extension of enhanced viewing for many people would be one of the CCD video camera systems offered today. These make it easily possible to share live viewing in our suburban surroundings with large groups of people, directly on a monitor and without the need for computers or software. I can attain a good image scale, equal to moderate (but not low!) powers, with a direct hookup to a CCD video camera. For a moderately high power, a combination with a 2× Barlow provides very good service. Everything will depend on the focal length of your telescope; my own telescope has a focal length of 81 inches and a focal ratio of F4.5. Telescopes with long focal ratios and much smaller apertures may produce a less satisfactory result. With such a camera connected to an image intensifier, the potential is far greater; now one can view or record deep space objects in true real time video. A complete system is also offered by Collins Electro Optics in conjunction with Adirondack Video Astronomy, who market a specialized video camera for the low light levels of astronomy, the Astrovid 2000. Nevertheless, despite the term CCD appearing in their names, the function of these devices should be considered in the realm of video; their simplicity of operation and moving video image keep them entirely separate from the standard astronomical CCD camera. For meteor showers, occultations, Jovian satellite transits, etc. the potential goes farther still, if this is your inclination. Video does indeed have some unique attributes.

If you are optically and mechanically handy, you may be able to adapt an intensified system other than the Collins to work with other CCD video cameras, particularly if you have already found a way to utilize an alternative image intensifier with your telescope. You will need, however, to incorporate the eye lens component into the optical path in order to produce a video image, since the intensifier alone does not have a "see-through" light path like a regular eyepiece. It will likely need to be placed further from the intensifier screen.

The real time video deep space images used in Chapters 7 were taken using the Astrovid 2000 system in combination with the Collins I_3 image intensifier (Figure 2.2). The original video footage was saved directly to hard drive via *Firewire*. The frames were selected and are reproduced here with minimal and

Astrovid 2000 CCD video camera

Collins camera-to-intensifier adapter
TeleVue/Collins eyepiece component
Collins extension adapter

Collins *I₃Piece* image intensifier

Figure 2.2. Image intensifier and video equipment mounted on the author's JMI NGT-18 telescope.

simplest processing (reflecting my bias towards live viewing), and provide a good overall impression of real time observations. I did not wish to further enhance the images beyond the resolutions here; the primary aim is to provide a guide for visual expectations, and not necessarily to extract the maximum detail beyond the image represented in the frame. You will find that moving video on the monitor, is generally more revealing than still frames; when contrasting video frames against drawings (also included wherever possible), you may likely be best prepared for what you may actually see. You may also get some idea just how hard it is to record the subtleties which only the eye detects, but for our typical suburban locations the examples may prepare you better than have many previously available materials.

For deep space video applications, a recursive frame averager (also available from Collins), becomes a virtual necessity. You will find that the effects of electronic noise from the image intensifier become quite pronounced on the monitor without such a device, and worse, result in unusable individual frames if you wish to record still images. Without a frame averager, you will find that as you work through the recorded video, frame by frame, you will be astounded by all the varieties of the same image you will see. This may cause you to wonder how all of them could come together to form a reasonably coherent moving image, to resemble a single deep space object. Frame averagers work by capturing a set number of video frames, and removing artifacts not common to each. The more frames in the sample (up to a maximum of 16 with the Collins unit), the more effective, technically, the device is, although the "softer" the image may be. Freeze frames may be utilized as well, in which multiple frames are averaged together to produce one clean still image. These indispensable devices are not inexpensive, but with no alternatives in sight, they are a part of the equation you must consider once you delve into deep space video with image intensifiers.

Just as in live viewing through the telescope, in video applications higher powers can readily be obtained by coupling the image intensifier to a Barlow lens. Not every deep space object will provide sufficient illumination for this always to be effective, though for Solar System viewing on the monitor without the intensifier (Figure 2.3), higher powers become something of a necessity. With my own tele-

Figure 2.3.
JMI NGT-18 telescope with complete Astrovid 2000 video assembly and video control box (on the table), which provides manual gain, shutter speed settings, as well as contrast.

scope's focal length of around 81 inches, a 2× or 3× Barlow, working through the already enhanced image size of the video image, is generally sufficient. (For deep space, very few objects are bright enough to support the use of higher powers than a 2× Barlow produces with my telescope's focal length.) Not every telescope has such a considerable focal length, of course, but here is where the technique of eyepiece projection will provide a ready solution. Such eyepiece adapters are widely available. Be aware that as the magnification goes up, so do the difficulties of locating objects, in addition to focusing issues; none of these is insurmountable, though.

Additionally, there is yet another way video imaging has great appeal; it is sometimes possible to see certain detail on the monitor other than what one is able to see through the eyepiece! On planets, the objectivity of being able to stand back and look at the image with both eyes seems to heighten the visual awareness, or at least it is a less straining experience. Astronomical video cameras often also respond favorably to the use

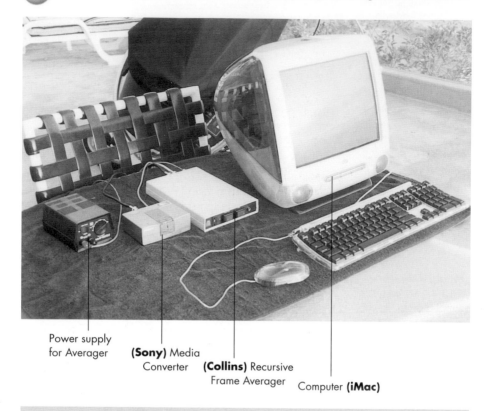

Power supply
for Averager **(Sony)** Media
Converter **(Collins)** Recursive
Frame Averager Computer **(iMac)**

Figure 2.4. Additional equipment used for recording images.

of standard color filters (although with monochrome cameras these results appear as contrast differences with a varying emphasis of planetary features), revealing details in much the same way as in normal live viewing. With the proper equipment, it is also possible to record the images directly to your video recorder as well as computer, as movie clips or still frames.

I once again must differentiate between video and CCD imaging. The latter is capable of producing some of the most stunning images imaginable, but remember they are not observable in real time. Video can indeed provide us with monitor viewing and a reliable record of something of the real time experience, often remarkably so. But there is still ample justification to draw what we see at the time, since there is still nothing yet to quite duplicate the way the eye registers detail. It also perceives momentary flashes of these details with subtleties and finesse altogether different to the finest CCD image. (Strangely, occasionally CCD, or even CCD video images can reveal certain detail the eye can miss.) These rare moments of clarity come about only moment to moment and no camera knows when these will occur. Here it is that the video camera has a unique strength: by examining what was recorded from the moving video record, frame by frame, it is possible to catch a particularly revealing frame of the object, when detail is suddenly in abundance, just as when the eye perceives those momentary flashes of detail. This simple method allows the user to store decent still images in a simple way, without the processing of more elaborate imaging systems.

To read further than this text is able to cover on the whole concept of using video cameras to view and record moving and still astronomical images, *Video Astronomy* by Steve Massey, Thomas A. Dobbins and Eric J. Douglass from Sky & Telescope's Observer's Guides series, provides probably the definitive information available today.

As far as time-exposure astrophotography and true CCD imaging are concerned, I must confess that as yet I have not spent time in pursuing these for myself. This is partly because so many fine images are readily available these days, and although it is possible to produce remarkable results even in light polluted skies, there is still no ignoring the fact that the resulting images do not provide a comparable experience to actually viewing their astronomical subjects live, the ongoing focus of this book. Because of this, I am not sure that I will ever muster the enthusiasm that would be required to produce the stunning results we frequently see these days with these techniques. However, if these other applications appeal to you, then I encourage you to pursue them with the enthusiasm they rightly deserve, even though the impetus of this writing is unlikely to be on the same road that you may find yourself traveling.

I should add that the above approaches are not the only way to produce exceptional camera images. Color images of surprisingly fine quality can be obtained with monochrome CCD video cameras taken through various color filters, later processed and combined on computer to produce fine full-color images of exceptional resolution. There are also color CCD video cameras available, such as the Astrovid PlanetCam. They are able to produce remarkably good planetary and lunar images in full color, live, and without the need for combining images taken through different filters, although they are not as light sensitive as monochrome versions. This disqualifies them from deep space applications, of course. Yet another way can be with a digital camera; it will also allow images to be downloaded directly to computer and enhanced in the processing. Coupled to an image intensifier and needing exposures of only a few seconds on deep space objects, digital cameras can produce remarkably well-resolved and illuminated images; examples, taken by Bill Collins on deep space subjects, may be seen later in this book.

Still another CCD video system comes yet closer to live deep space viewing, but there are major differences. Aside from live video, this deluxe system (the STV by SBIG) allows the user to build up images in single video frame exposures (up to 10 minutes) and witness them on the built-in monitor, or external monitor. It is also possible to combine multiple exposures and view them as they are accumulated. Not quite the same thing as actual real time viewing, but I think you'll agree it may be a step in the right direction, and a viable option for many. The results are apparently remarkably good. Recently, a new Astrovid video camera was introduced – the StellaCam. This, along with its more sensitive counterpart, the StellaCam EX is also capable of producing live deep space views of the brighter objects on a monitor, and the less bright ones by accumulating frames in a similar simulation of real time as the SBIG system, but at a significantly lower cost. Certainly, priced at US$695 (2002), it is the amongst the most affordable options for the suburban astronomer. It is claimed that the StellaCam can effectively increase the aperture of any given telescope by two to three times, a similar claim we already know about image intensifiers, and a key issue for us in the suburbs. (It can also be attached to the Collins I_3; the results would be interesting at the very least, but as yet, I have not tried it.) The manufacturer, however, does state that its resolution is not the equal of its top-of-the-line Astrovid 2000, and one should bear in mind that with any camera one is still not looking directly through the telescope, but indirectly at a monitor. Nevertheless, if budgetary constraints make the purchase of some kind of effective image intensifier out of the question, then maybe one of these systems offers a good option, and the essence of this book will still remain largely valid for such an approach. It has to be cautioned, though, that any of these video systems may be a whole lot more trouble to set up and use for live observing sessions than you may be accustomed to.

Additional Equipment Suggestions

Another great value and a source of remarkable ease in observing would be the addition of "digital setting circles", with many varieties available for most models of telescope, even non-equatorials, and featuring extensive numbers of programmed presets. They make the rapid sighting of objects possible with a minimum effort, and with surprising accuracy. For suburban viewing, paying for tens of thousands of programmed objects would seem to be largely an exercise in overkill, since most of these will be below our threshold of visibility, no matter what provisions we are taking to combat viewing problems. However, assuming you plan to visit dark sky sights at least once in a while, it makes sense to pay for a good base of, say, 2,000–12,000 objects in the preset memory. The addition of this accessory seems inexpensive when you consider the great facility it will provide. Don't listen to those fanatics who insist that you must know the sky intimately and "star-hop" your way around for object location. Not too many would-be enthusiasts will stay the course; there is no knowing how many have been lost in the past due to this type of elitist thinking. Digital circles are doubly valuable to us in the light skies of the suburbs, where a dependence on "star-hopping" may also involve searching for near invisible signposts. In a similar vein to digital circles, I really do not see the necessity, other than a sheer fascination for robotics, of the so-called "Go-To" scopes. While it must be very novel to have the telescope automatically find and slew to any object, you will be paying considerably more for a function that can easily and quickly be carried out by hand. Using digital circles, with numeric pointers to each location on the read-out, it certainly doesn't take any longer.

No discussion of viewing aids would be complete without some reference to standard color filters, which are often well suited to our planetary viewing. Do not expect more than certain subtle enhancements from them, however. The benefits have often been overstated by those in the business of manufacturing and selling them. Buy the most appropriate ones for your telescope's specific aperture; for this, you should consult your dealer, since the darker filters will only provide satisfactory results with larger apertures, and may result in obliteration of the subject in too small an instrument. I find that once I have details firmly in my eye and mind from having used filters, it becomes easier to make out the same features without such filters. (Another example of "educating" the eye.) My planetary drawings typically represent this final stage of the observation process.

If you use any type of telescope requiring you to look upwards through it (such as refractors and catadioptrics), star diagonals become quite essential for objects high overhead, but be aware that they also laterally reverse the image as well. This further complicates matters if utilized in any way for video or other imaging.

For the type of observing emphasized in this book, the Moon and the primary telescopic planets very much suit our conditions in suburbia, since they present images generally not impacted by light pollution. Together with the fact that they are such a prominent part of the history of amateur astronomy, they will always feature prominently in any discussion of real time observing. And by the way: do not even think of applying an image intensifier to these bright and close subjects! This will quickly result in ruination of a very expensive accessory. Again, image intensifiers are for faint, usually deep space objects. I would encourage you, though, to try your image intensifier on Neptune, and especially Pluto. On the latter, it will prove an invaluable aid in seeing this lonely outpost with something approaching ease, once you know where to look. Uranus is a borderline case, and may be too bright for the intensifier through larger instruments. Beware!

A great bonus that comes with the Solar System is the rare opportunity it provides to see actual color in outer space with our own eyes, not including colorations of individual stars, of course. Because of the many changes continually taking place on these other worlds, they give us a never ending reason to view them, and I have found that I tend to become totally absorbed in any given planet when it is near opposition, often to the exclusion of most other observing. Of course, when it comes to deep space, the kind of fleeting and fine detail we associate with Solar System subjects is not usually in the cards anyway in real time viewing. Aperture now becomes critical, just so the object can be seen at all! The air need not be so steady, and lower powers are often the most effective, except in cases where we can use higher magnifications to darken the background sky. We are unlikely to see much in the way of color in deep space; with an image intensifier the issue is moot anyway. Hopefully, we will all be fortunate enough to have access to the appropriate equipment to take full advantage of every type of astronomical viewing available to us as real time observers.

Chapter 3
Techniques for Seeing

There is infinitely more to seeing through a telescope than merely pointing and looking. Smaller telescopes are often discarded way too soon because their owners fail to recognize the full extent of what is present in the image. Often larger apertures will not satisfy these same owners' hopes either. In our suburban locations, the issue becomes even more important, since we will be trying to extract the maximum possible from less than ideal circumstances. For each type of observation, be it lunar, planetary, sights within our own galaxy, or other galaxies, there is an optimal method of seeing all that is there, and in many cases, learning to draw what we see is the best training possible to accomplish this. Beyond that, of course, there are issues of equipment and conditions to consider, particularly if we are trying to overcome in any way the negative factors involved in suburban viewing. However, don't only rely on or blame these for your success or failure, as there is much you can do yourself to maximize your viewing. Then there is another factor entirely: knowing what sights are best suited to our conditions and the specific approaches we can take with each. Let's review some of these.

The Moon

This is a natural for us (Figure 3.1); even with the worst case of suburban light pollution, the Moon will be

Figure 3.1. The Moon.

totally unaffected. So many books on lunar observations have been published and are readily available that it becomes redundant to weigh down this writing with endless maps and surface descriptions. More in Chapter 5.

Because the Moon is so bright you will find that with larger apertures some kind of filter will be needed as the lunar disk grows from crescent to full. These are commonly available and inexpensive, of different types and by numerous manufacturers. Without one, however, you may not be initially aware of the full extent of the Moon's dazzle. Only when you step away from the eyepiece will you realize just how much light has been pounding on your retina! It can affect your vision for longer than may be comfortable. While I don't know of any specific cases of actual damage to the eye, common sense tells us that as an extended practice it can't be doing your eyes any good. All that a moon filter will do is reduce the total illumination, but not at the expense of detail, or the addition of false color. Because it is so bright, the Moon is able to withstand some of your highest magnifications, but generally you will find that something less than all out will

give the best results. It is with these moderate powers
that you will have the best illumination, crispness and
contrast. There is no doubt though, that during nights
of good "seeing" the Moon will reveal a maximum of
detail with somewhat higher powers, and certainly
sends us so much light that using them whenever possible is not too much to be frowned upon. In Chapter 5
I have featured some images taken in real time through
my telescope via video camera that will give you a very
real sense of what you can expect to see, although I
know of no means available to us that can really convey
how amazing the Moon appears live, through any telescope. It seems to take on an almost three-dimensional
effect that does not fully translate to any form of
imaging.

The Planets

Again, these objects (Figure 3.2) provide ready access
to us from anywhere near the city. They are unaffected
by light pollution, and may even benefit from atmospheric impurities, which can act as filters. Some of my
best views have been through anything but transparent
air! Above all, we are looking for steady air when
viewing the planets. Throwing the image out of focus
will immediately reveal just what we are dealing with at
any given time. When there appears to be a current of
air in one continuous direction across the out-of-focus
disk, "seeing" is likely to be at its absolute worst. Look
for slow undulations in the out-of-focus image, slow
mixed-up motions, or ideally no movement at all
(right!) for your best viewing. In the meantime, unless
your telescope is out of optical alignment, or of poor
quality, do not blame it for blurry views when conditions are not right. At these times even the best telescope, and all the more so with increasing aperture,
will perform like a very inferior instrument. It is also
true that with telescopes of short focal ratios, the need
for today's high-tech eyepieces is a near requirement
for planetary viewing; without them, you are likely to
blame the atmosphere, or worse still, your telescope.

When it comes to viewing the Moon and planets,
much has been said about the value of "stopping
down" the effective aperture of larger telescopes during
times of more turbulent seeing. The value of this, particularly when utilizing offset masking, is the prospect

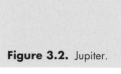

Figure 3.2. Jupiter.

of producing an unobstructed primary objective in a reflector. With my telescope, it is possible to produce an unobstructed aperture of 6 inches; such an offset mask adjacent to the primary allows the surviving light cone to avoid the secondary mirror and its supporting spider stalks during its entire travel, an interesting prospect to be sure. However I have never felt that this brings about much advantage, and cannot recommend it; with such a reduction in effective aperture, the scope seems to lose more than it gains. While being aware of the theory behind offset masks, I do not agree that they produce the best results in poor conditions which are nevertheless still passable for planetary observing. Images at full aperture usually remain preferable to this writer, even though moments of good seeing will be more fleeting. Simply stopping a reflector's primary down to a lesser effective aperture is of no value, since it renders the secondary much too large in relation to the primary.

Even if the planets are not your primary interest, though they have always been of paramount interest to me, they can teach us to see. Additionally, they can teach us to draw what we see with the greatest chance of something approximating completeness, as they are so much closer, brighter and more readily apparent than deep space objects. The main advantages of large apertures are less significant when it comes to the Solar System; telescopes larger than 10 inches in aperture rarely achieve their theoretical resolution, owing to atmospheric instability. This means that most of the detail that could be seen with these larger sizes simply will not show under most circumstances, and the main benefit will only be brighter and more color-saturated images. This is not to say that these larger sizes are wasted; all observers would agree on the relative viewing luxury that they bring on a good night, but the best of the moderate sizes will reveal most of the visible detail.

Remember, you will be spending long sessions at the eyepiece trying to discern exactly what it is you are seeing. Although these details may be quite clear when they reveal themselves, they are typically hard to "get a handle" on when it comes to drawing or describing. Regularly throw the image out of focus and readjust. You will find the strain on the eye is greatly reduced and the newly focused image seems all the sharper. Most of the time you will be wading through unsteady or blurry views, with detail suddenly flashing out with amazing resolution. You have to train yourself to snatch all these moments and try to comprehend what you are seeing during those times. Very occasionally, "seeing" is so good and steady that planetary detail stays sharp and defined for prolonged periods. The luxury of actually observing in such conditions cannot be overstated. The rest of the time we have to learn to make the most of what we have. Motorized focus can also assist you; it completely removes the tremor of hand focusing in the slight increments at the higher powers that planetary viewing requires. Such devices are readily available, relatively inexpensively, for retrofitting on most telescopes. JMI is famous for them.

During these extended sessions, the need for equatorial or other excellent tracking capabilities becomes all the more important. This is not to say that I always had them, and good results can certainly be enjoyed with a lot less. However, these days I have come to regard good tracking almost as a necessity; it has become an indispensable aid to countering external frustration during my own observing. Toward this end, if your telescope is equatorially mounted, it is highly advisable to set it up to be as accurately polar-aligned as possible. It makes all the difference for satisfactory viewing; while five or more minutes may seem a long time between adjustments, in planetary viewing sessions, five minutes flashes by like five seconds. If you are at all like me, you will find the constant "tweaking" to maintain a centered image becomes a real nuisance and handicap, once the novelty of seeing well has worn off.

The planets, like the Moon, will also withstand considerable magnification. Just how much depends on each particular planet, the telescope, and the air. Generally, air turbulence of one form or another will place an upper limit on this, regardless of aperture. Normally, powers above 350× require excellent conditions, no matter what telescope you are using, but smaller apertures will give more per inch of aperture up to this point. It is usually unwise to try to reach the often stated limit of 50–60× per inch. This power is reliable only for apertures up to 6 or 8 inches, and usually does not refer to all types of viewing. Double star separations or other such subjects, not requiring the finest clarity of detail, are the usual beneficiaries of this kind of power. You should always try to look for maximum resolution with the lowest power necessary to bring it out. I will admit to very occasionally pushing the power to the upper limits in great conditions (i.e. 600–700×) when trying to discern detail, say, on one of Jupiter's moons, but almost all of the time, much lower powers than these bring about better results. Remember that lower powers produce easier tracking with generally brighter, more contrasted images and impressions of color.

Star Clusters

Star clusters, particularly the globular variety (Figure 3.3), are relatively well suited to our suburban conditions. You will find low-to-moderate magnifications are best for most of your viewing, and with reasonable aperture they are wonderful sights; some globulars can withstand much higher power. Many can even put on a good display without any enhancing devices at all, and are the most beautiful this way in some respects. However, they will be at their most striking with image intensification, being congregates of points of light. Here, there is nothing diffuse; either a star is visible or it isn't. With intensified star images though, the brightest appear the largest. In normal viewing, the brighter globulars share a uniform characteristic of appearing something like a large glowing Christmas tree ornament, and seem more spatial than just the two dimensions

Figure 3.3. M13 in Hercules, intensified image. Courtesy W.J. Collins.

they present to us. Unfortunately, when observing them with an image intensifier, this characteristic is lost. However, just as with open clusters, they are most likely to be resolved into multitudes of bright stellar points that will dazzle even the most jaded viewer. The skies contain many fine examples that are comfortably within reach from the suburbs. Intensifiers also make dark lanes and other interstellar matter quite conspicuous in these objects, to a degree that would have startled observers of long ago, who strained to see these features in much darker sky conditions. When looking for such lanes, note the extremes of the cluster's dimensions, or even resolution of fainter stars, and you will find indirect sight does make a difference, together with the use of high powers on the brighter ones.

Those clusters best suited to our purpose would seem to be where the entire cluster, or most of it, can be easily contained within the field of view, and also where there are sufficient stars in the cluster (at least in the magnified field of the telescope) that the view is reasonably saturated. Many of the open clusters are far more suited to binocular viewing or the low powers of "richest field" telescopes, and will not be more than casually dealt with here.

Planetary Nebulae

Planetary nebulae are also wonderful targets for the city observer. Those that respond to intensification do so astoundingly, and it is typical that their often elusive central stars become easy sights. Even the famous and usually invisible central star of the Ring Nebula M57 (Figure 3.4) becomes almost conspicuous on a good night with larger apertures, in spite of its light being in the blue part of the spectrum. It is interesting that some planetaries do not seem to respond at all well to intensifiers, and may often be finely enhanced by a narrowband light filter instead. This is partly attributable to the nature of light being emitted by the specific nebula, and is also, apparently, affected in the suburbs by its size as well as magnitude. The smaller ones that are listed in Chapter 7 are quite bright when viewed with the Collins I_3, the total brightness being spread over a small area and able to withstand relatively high powers. (All of them will benefit from experimentation in this regard.) The larger planetaries, while maybe having an encouraging total magnitude, are more often quite faint visually, and tend to be less likely to respond favorably to intensification.

In our suburban environment, you will find that, as with other vaguer subjects, planetaries tend to live up to what we might call the magnitude 12 rule – that is, anything fainter tends to disappoint, or not be visible

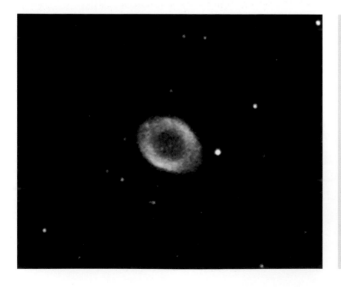

Figure 3.4. M57 in Lyra, intensified image. Courtesy W.J. Collins.

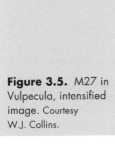

Figure 3.5. M27 in Vulpecula, intensified image. Courtesy W.J. Collins.

at all. In light polluted skies, good candidates for the use of light filters instead of image intensifiers would include the large but bright Dumbbell Nebula M27 (Figure 3.5), or the Little Dumbbell Nebula M76. Other famous planetaries, such as the Owl Nebula M97, do not seem to respond well to any method from the suburbs. You will find, however, that most planetaries are small, with some notable exceptions. With light filters and even low powers, their frequent bluish glow makes them easy to separate from the surrounding stars. I do not find averted vision necessarily beneficial with most planetaries, except in cases where their dimensions are large enough that mottling or other subtle shadings can be seen. Those smaller ones which are easily visible tend to be straightforward to view, and reveal their natures readily.

Emission and Reflection Nebulae

Emission nebulae, fluorescing from the radiation of an embedded star or stars, such as the Great Nebula in Orion M42, can be great targets for either filtered or intensified viewing. Greater challenges in the vicinity of cities are reflection nebulae. Viewing them successfully depends on the type of light being reflected from the illuminating star(s). Some nebulae are not exclusively emission or reflection varieties, and they can therefore

Figure 3.6. M17 in Sagittarius, intensified image. Courtesy W.J. Collins.

benefit from different approaches. Surprisingly, there are a decent number of nebulae to see in the suburbs, and I have found that I can gain access to different aspects of some of them by using alternately a narrow-band filter, no filter, low and high magnifications, and in many cases, image intensifier. The Omega Nebula M17 (Figure 3.6) and the Lagoon Nebula M8 are such cases in point. All four types of viewing produce stunning results, and provide different insights. Averted vision can be very helpful in discerning detail in these filamentary structures.

Galaxies

We must be realistic here. There are limited opportunities to see spiral arms or other structural details in real time from suburban locations, though it is by no means impossible. A lot will depend on the clarity of the air, the degree of light pollution, the particular galaxy, as well as the aperture of the telescope we are using. More to the point: even when spiral resolution is not possible we can nevertheless see enough detail or primary features to make this form of observation perhaps the most interesting of all.

In traditional viewing, the best chances for seeing spiral structures have been with face-on spiral galaxies. Because their full form is directed towards us, these galaxies would still seem to offer the greatest potential of all, but are the same ones that do not respond well to image intensifiers, and present significant viewing problems in the suburbs. Quite a contradiction, it would seem! Image intensifiers aren't especially sensitive to the frequencies of light (much of it toward blue wavelengths) that emit from these spirals' arms, which are being seen directly from above or below. We are most likely only to see an increased brightness of the central core, where more favorable light wavelengths are generated, further detracting from any impression of a spiral. But all is not lost. Some of these same large face-on galaxies respond to narrowband light filters, and on occasion I have seen something of the spiral arms of M33, M51 and others in my light polluted neighborhood when the skies are clean. Low-to-moderate powers will give the best results most of the time, though higher powers can work when the subjects allow, and the image responds to the increased darkening of the background sky. It is also possible on occasion to see nebulous emission regions in these galactic structures, not unlike the Great Nebula in Orion M42 in our own Milky Way. These galactic nebulae will look only like little bright patches from these distances, but it is truly mind-boggling to see something such as these across time and space. Some of the globular star clusters that exist as well around those galaxies, such as within the Great Andromeda Galaxy M31, remain difficult objects, but they are not impossible with image intensifiers and sufficient aperture in suburban conditions. However, don't expect to see them as much more than points; by carefully referencing detailed charts it is not too difficult to know what you are looking at. Such searches have become an obsession to some.

One of the biggest challenges with these large, close, face-on spirals is to subdue the anticipation of seeing a brilliantly illuminated galaxy. Some of the most celebrated spirals have great dimensions and total magnitudes; these magnitudes refer to the object in its entirety. Such a total magnitude could also refer to a single star point. Increasing size and total magnitude do not usually mix favorably, as image brightness is diluted by dimensions. With the larger ones in particular, their bright total magnitudes are not indicative of how they will appear in the eyepiece. People frequently miss these objects completely when looking for them, simply because they are not expecting anything as extended or dim as so many appear, and the observer may have observatory images more in mind. It takes long periods of time with trained eyes, searching for order out of confusion, before these galaxies' structures begin to become clear. And they still will not look like astronomy book pictures!

Edge-on and near edge-on galaxies are another matter entirely. From this viewpoint, the red and infrared spectra that become increasingly dominant with them make for much more exciting results when we use an image intensifier. Those with equatorially encircling dust lanes can produce truly astounding views, even in the relatively troublesome skies of the suburbs. The best examples do indeed closely resemble their well-known photographic portraits, once the eye has settled down to the reduced contrast and brilliance of the live image. NGC 4565 and NGC 4594 (M104) are two of the most breathtaking examples. Elliptical galaxies also send us a very usable spectrum, regardless of their spatial attitudes relative to us, and usually show greatly increased brightness with an intensifier in a suburban sky. Visually, though, these tend to be the least interesting galaxies. In most cases, irregular galaxies are likely to be too diffuse to show well in the suburbs, though M82 (Figure 3.7) is probably the most famous exception.

However, it is also true in the suburbs that a significant effect on all of the above is the apparent

Figure 3.7. M82 in Ursa Major, intensified image. Courtesy W.J. Collins.

magnitude of the galaxy itself. High magnifications can produce good results on some of them, although I have found that there is no hard and fast rule. However, here again the magnitude rule applies: 12th magnitude galaxies, and fainter, are most likely to be disappointing or difficult to detect in the suburbs, even with the maximum aperture available to us. For fainter sights than these, we will probably need to relocate to a better site, as nothing I know of can defeat the ultimate limitations posed by city dwelling. This is not to say that we should despair, as there is still a substantial array of galaxies available to us, and quite spectacularly so.

In normal viewing, the use of increased magnification will help darken the background sky, throwing the subject into better apparent contrast. The same approach will be even more helpful with an image intensifier. Bear in mind that we are amplifying all light, and city glow is no exception; the background sky will also take on some green hue, depending on the amount of light pollution present. Intensifiers would not be at all suitable for our purposes if they required us to use extreme powers to overcome such background glow. Luckily, this is not the case, and a modest amount of power increase, such as twice, with the use of a Barlow lens, will more than likely darken the background sufficiently. Without so doing, the sky will probably appear distractingly green, and bright at that. Of course, increased skyglow, humidity and particles suspended in the air all snare light, and will all affect the performance of the image intensifier. You will still treasure the best nights at your observing site.

In addition to making the image much more pleasing, boosting the power with a Barlow will also make the intensified image easier to bring to a position of focus, for straight-through live viewing in most telescopes. Certainly focus issues are frequent comments about the Collins intensifier, which has been known to change this position radically, to the point of creating difficulties in finding focus at all with some telescopes. The Barlow lens, though, will usually fix it in every instance. If you are at all concerned, however, check with the manufacturer, since it is likely that the device can be supplied in a configuration specifically for your particular telescope. The magnifications I commonly use with it, in combination with the 2× TeleVue Barlow lens, are approximately160×, and with a 3× Orion Barlow lens, 240×. I also use my standard low power eyepiece, equipped with narrowband filter to locate and center the object I want to observe, only then switching to the higher powered intensifier. This procedure will save you time, frustration, and wasted wear of the intensifier, as well as any potential risk from snaring a nearby bright object. Interestingly enough, focus issues aside, I find that the above benefit of combining a Barlow lens and the image intensifier together to reduce skyglow is a benefit with visual use only. If you should try your hand at intensified video imaging (see Chapter 4), none of this applies, the background sky usually appearing normally illuminated. The Barlow may simply be used as a means of expanding the range of power, as in normal practice. You will find, however, that in video applications, any use of Barlows in conjunction with an intensifier will probably only be successful with the brightest objects; the rest disappear into darkness.

Recent advances, previously mentioned, in light filters made specially for image intensifiers include those offered by Collins Electro Optics. One such filter is the IR band-pass filter, which eliminates sodium and mercury vapor light pollution wavelengths while allowing infrared wavelengths to pass through. It would seem that this is a further enhancement of the solution to our situation! However, I have not had good success with these filters as of this time, and cannot attest to their effectiveness. As explained to me by Bill Collins, this is presumably because of the high humidity of my particular location, which cancels out much of the infrared frequencies of light able to penetrate the local air. Users in arid areas would probably benefit from the filter's use, as their application is apparently very atmosphere specific.

A word of caution! You will also have to learn the art of "seeing" when using an image intensifier. In no way is all that it is possible to discern immediately obvious with these devices, although I will readily grant that the task is made much easier when using one. Nevertheless, I had to learn the art of intensified viewing over a period of time; now it seems amazing to me the number of significant sights I had originally dismissed early on because I was unprepared for this specialized type of viewing.

Using regular light pollution filters for standard eyepieces is a simple topic. You simply screw them into place on the eyepiece and return to viewing! Remember that the specific type of observation you are undertaking at the time is everything when it comes to judging their effectiveness. It is not realistic to expect any of these devices to do what they can't. Some of the specialized filters such as Oxygen III and H-beta are designed for a relatively limited number of deep space subjects, such as the notoriously difficult Horsehead Nebula. Many people report that the advantage is not much more than with the best of the narrowband variety. None of these filters work beyond their unique capabilities, and none can duplicate the environment of a true dark sky.

So, in conclusion, while it is indeed quite possible to see or glimpse many of the sights listed and described in these pages without any means other than a moderate-size telescope, if you want to experience a true revelation in suburban astronomy you will need to go further. You should find much in this book that will guide you in deep space, regardless of your own access to such equipment, or even your location. However, for the truest intentions of this writing, the acquisition of anti-light pollution accessories will make the quest far more in line with your original hopes and aspirations. Image intensifiers and light pollution filters provide the greatest potential to drastically enhance your viewing, with intensifiers being the most significant of the two. However, in no way would I want to imply that of the two types of device, light pollution filters are a poor second-best. Each one fulfills a different role, and highly important individual factors in the quest to widen our access to the night sky. If financial considerations preclude more than the purchase of just a narrowband filter (the least expensive option), you will still see much more than might otherwise be the case.

Chapter 4
Introduction to Drawings and Real Time Video

In learning to be aware of all that is present in the eye-piece image, the value of drawing cannot be overestimated, and it should be regarded as more than a simple and ready means to record images. The process of drawing will reveal to you far more than you otherwise might see. I cannot pretend that any of the drawings I undertake have much scientific value; the technologies available to modern observers and observatories, together with specialized space reconnaissance, have accessed the universe in ways that we could never have dreamed of just a few years ago. However, even with the relatively modest equipment that most amateurs have, seeing and recording what can be seen for oneself remains unique territory, and highly worthwhile on a personal level. While I do not pretend to draw what I have seen with exact geographical precision, I do try to document how I saw it, and to match closely the visual impression I experienced at the time. Sometimes I am successful, other times not. In striving to accomplish this goal, I find my eyes stretched to new limits, and also interestingly enough, my visual judgments. Those who have not tried to do this will never fully see all that is there, and the images they may even record by CCD or camera, no matter how good, still do not reproduce quite the visual impact they might have experienced as real time observers. Maybe a lack of awareness of the value of this fundamental root of astronomical observing could explain how it is that there are many who seem to show no enthusiasm for real time viewing. Perhaps, in a world of instant gratification, their own experiences have proved to be disappointments.

Instead, they have become, in effect, remote-controllers of telescopes, expert operators of equipment.

It used to be that the areas available to the amateur to do useful work were quite wide and varied. From the dawn of the space age this has slowly but surely been whittled away, although there are still some limited areas available to us if we wish. This is not the purpose of this book, however, and we will not therefore be setting out to do what we could have readily engaged in just a few years ago. But is there no virtue in being enthusiastic sightseers in the universe? I would certainly say there is, just as the value of travel and seeing different places around the world is never questioned. So, let us not belittle the sheer enjoyment of fine astronomical views and the enhanced knowledge and awareness it brings, just because it may not involve actual contributions to scientific research.

In recording by some method what we have seen, it is also most satisfying to review these sights over the years. This is particularly so if we are trying to capture true visual impressions of what we have seen, as opposed to documenting in some way what was seen, without portraying much of the actual visual effect. In certain instances, I still know of no better way to attain this goal except by drawing, since the subtleties that separate it from other methods are frequently hard to record by any other means; it seems that this low-tech approach may still have its place in our high-tech age. It is also wonderful to have some kind of record-keeping of our efforts as we refine the process. Some images will stand out as memorable, not only for what we saw but also for our success in transferring those sights to paper. Realize that I am stressing that there is a real difference between recording in some way the essential features of what we have seen, and something that attempts to recreate the visual experience. The latter is always my objective.

Space objects are so frequently drawn with a lack of realism that I thought it worthwhile to share all that I could impart to the process. Specific methods are so rarely described, that maybe you will indulge me the considerable detail of my own drawing processes that I outline in this writing. Hopefully, you will enjoy similar or better results, and you need not be a second Michelangelo. Here are some broad principles, in preparation for the successful drawing of deep space objects at the telescope.

Most people draw things with too much contrast and a sense of stiffness that does not equate with how the object really looked. In reality things are much more subtle and fluid.

Deep space objects are in some ways the easiest things to draw. With few exceptions, you will never need to use color, as the objects appear in faint shades of mostly white or gray. With the use of an image intensifier, this is not a factor anyway. On the sketch book page, deep space objects appear to me essentially as they did, when simply portrayed as negative images in exact reverse – that is, black on white, using standard lead pencil on white paper. They are certainly easier to produce than the other way around, appearing quite lifelike, without diluting the effect of your recollection of the objects' appearance in the eyepiece. These images are perfectly suited to your observing sketchbook. (It is simply too difficult to try to draw deep space objects with white pastel pencils on black paper; that medium is not flexible or subtle enough to produce the quality of image that standard lead pencil on white paper so readily provides us.)

To go a step further: if we are able to draw these deep space images with the realism of reversed light and shade, the potential of a finely executed image is greatly enhanced when scanned into a computer. A successfully made drawing (Figure 4.1a) will lend itself readily to reversing from black-on-white to white-on-black (Figure 4.1b). (The additional mirrored reversal of the illustrations here represents a little artistic license only!) With just a little adjustment of contrast and brightness after reversal, this is a sure test of how well the object has been drawn on the page. The areas of luminosity will seem to stand out real as on a photograph, and the gradual blend of many objects into the dark background will transfer effectively. Some results obtained in this way astound me. Conversely, less than well-executed drawings will immediately reveal their shortcomings; it is an easy test of your skills.

When drawing any deep space object, it is worthwhile to give some consideration to any significant surrounding reference stars. If not immediately obvious to anyone looking at your drawing, such reference points immediately provide orientation when comparing the drawing to known images, and also help to instill a sense of how the object looked in its own field. Individual stars' brightnesses can be well represented

a

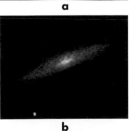

b

Figure 4.1. **a** Black on white drawing, and **b** its mirrored reversal.

by varying the size of stellar dots, much like their appearance through an image intensifier.

Because of the problems of being sure of the actual true orientations of new and unfamiliar objects with the rotating-eyepiece positions of many Newtonian reflectors, lateral reversals of star diagonals, etc., I will not confuse things further by indicating on the deep space illustrations the positions on the celestial sphere. Remember image intensifiers also project an upright image, so this adds an additional complication; be aware of this when consulting star charts. I have decided simply to present the deep space images as they appeared to me at the time of viewing. This should be simple enough to interact with your own viewing, and known images.

For drawing, standard erasers as supplied on most pencils are not well suited to our purposes. Equip each pencil you use with a flat "inverted V" style eraser. These erasers fit on the end of the pencil, and because they gradually taper to produce this V shape, it makes a substantial difference in accurately shaping what you are transferring to paper. Also buy a pencil-type eraser for type-print, and sharpen it to a dull point; this will prove enormously helpful in removing dark streaks and spots that sometimes appear when the paper does not take the pencil evenly. The hard point of this type of eraser is perfect for dealing with these.

More specific principles for drawing deep space objects:

- Use a soft lead. Shade in the general features of what you are drawing, but only very lightly. Remember, most people exaggerate contrast; less is always more for accuracy of appearance in astronomical drawings. By smudging and blending with fingertip and eraser, the general shape and features can be reliably shaped and placed. From this point, it is best to apply darkening by carefully "dotting" with the pencil point, until the desired depth is achieved. Then blend again with finger and eraser. Anything else tends to be too blatant, detracting from the realism; few deep space objects are so clean cut in appearance.

- Blending and smudging the margins, particularly with galaxies, will greatly enhance the realism. When your drawing is advanced sufficiently, leave it and return shortly after to reappraise if it measures up to what you saw at the eyepiece. Seeing it afresh this way several times will enhance your objectivity. Continue blending, shading, and lightening until you are satisfied that you have indeed captured a likeness. Often, just the slightest touch, in the wrong place, or not correctly blended, will make a radical difference in capturing the "look" of the real thing. In this way I often return to the eyepiece and compare the effect of what I see with what I have drawn, long after I may think it is finished. It is surprising how much I will keep tinkering with my drawings until I am satisfied I have recorded as accurately as I can the way the object came across to me.

- Remember that the eye needs whatever amount of dark adaptation it can have. This is more critical for deep space than anything else, although image intensifiers diminish the depth of it that you can attain. When drawing at the telescope you will need something resembling a clipboard, preferably with a red light attached. Something else when using image intensifiers: even the soft glow of a red light, and pointed away at that, will tend to be picked up in the field of view! Sometimes this is glaring. I have found it necessary to alternate use of the light with the intensifier when making drawings, which does make the process more tedious and trying. However, a small table or stand right next to the telescope is a great assistance; hands-free time at the eyepiece will help during these times.

Returning now to the planets: these present some great opportunities for us, and also some greater drawing challenges, to which some additional general principles apply. Here we can put our eyes to work in ways that will only enhance all other types of viewing, as well as record some real time imagery as only the live experience presents to us. For specific planets, I will return to them with more information in Chapter 6. For successful suburban viewing, they do not suffer from the degrading effects of light pollution, and can do quite well even from low-lying locations. To see them at their best, however, we must recognize that such low elevations will rarely present us with the atmospheric steadiness of higher altitudes. Therefore, we need to take advantage of every moment of steady seeing we have, and if we do, there is no reason our results cannot be first class.

With the planets, it is wise to bear a few things in mind. Planets are rotating spheres. There is nothing

unusual in the universe about that, except these rotations take place in a time frame we can actually appreciate. In the case of Jupiter and Saturn, they are
essentially gas worlds and reveal the effects of rotation
on their features and formation. This is something to
remember as we draw them, as all features on these
spheres must have some sense of rotational relativity to
each other. Often there exists on amateur drawings a
lack of connection between one part of the disk and
another. If in doubt, compare one of your early efforts
with a fine CCD or spacecraft image. I am sure you will
see quite easily what I mean.

Drawing the planets will be the best opportunity we
will ever have to record color in live viewing of outer
space. However, on all planetary subjects, take care to
represent subtleties of color accurately; these are not as
extreme as they may seem after a long session at the
eyepiece. CCD photographers frequently exaggerate
colors in the processing in order to make the features
stand out. Do not necessarily be guided by the colors
on these images. Always color in the features slowly
and faintly at first, gradually deepening them to the
point where they can stand the scrutiny of an objective
reappraisal.

As with everything, in the case of the planets it is
always the appearance in the eyepiece that I try to
appreciate and draw. This drawing process actually
becomes a two-stage affair. At the eyepiece, I try to
record everything in sketch form with lead pencil, with
descriptions, detailed as necessary, for most features.
All of this will strengthen my memory when I go
indoors to produce the final drawing in color. It is wise,
when learning the drawing process, to again revisit the
eyepiece after completing the drawing to make sure
that the contrast effects are realistic. It does not matter
that the planet will have undergone additional rotation,
as we are primarily interested in its overall visual
appearance. Always remember that most planetary
drawings suffer from excessive contrast and color.
(Even the contrasts of the reproductions in this book
become somewhat exaggerated in this respect, and
lacking in the subtleties of the original.) This is the
natural consequence of a second or third generation
image. Over a period of viewing time, the mind does
seem to suggest greater color effects than are actually
there. In making the final drawing, the mind's natural
sense of visual clarity can lead to exaggerated
renditions.

Unlike the deep space images, I have presented all of the planetary images with south up, in order to coincide with traditional practice. (The lunar images in Chapter 5 are not orientation-specific.) Once the planetary image is in the field it becomes very easy to adapt the brain to a consistent orientation, owing to the great familiarity of these subjects. It is normal for any planetary observer always to orient the eye with south up, and to impose that visual logic on all planetary disks. This is a familiarity which is acquired early on in one's astronomical pursuits. The time we may have spent drawing deep space objects trains us well for drawing the planets, which are considerably more complex in their fine detail and are in full color.

Specific techniques I use for capturing the essential "appearance" of planetary features:

- Use blank disks of the appropriate size and shape (see Chapter 6) and fill in the surface features. Just as you did with standard lead pencils for deep space objects, equip each color pencil you will use with the same flat "inverted V" style erasers. After lightly outlining the basic features in an appropriate color base (usually gray), use a fingertip to spread this out evenly. All the pencils I use respond very favorably to this method. Lighter areas on the drawing that become darkened, or that have changed shape on the page inadvertently, can be trimmed into shape by the erasers installed on any given pencil. Where sharply realized definition is required, unworn eraser corners will be found very suitable.

- You only have to make one drawing to see immediately why it is so important to have an eraser on each pencil. This is something which, incidentally, distributes wear amongst many erasers, which is more of an important consideration than may be immediately obvious. It is rare for these erasers to be used up; they will likely have to be changed fairly early since their shapes will deteriorate beyond usefulness. You will find them indispensable tools for forming your images. Lightly dabbing them directly on the pencil applications will blend the outlines and give you control for producing the subtle effects we strive for. As with deep space objects, I find as the drawing proceeds sometimes applying these pencils in dots often gives me the control of demanding shading. Following this with the blending process of finger and eraser, I can usually apply the color closely matched to the subtleties of contrast that I see.

- Always use a light motion and touch, building color and subtleties of the features gradually. Small motions will serve you well, with constant changes of direction. This is often best accomplished simply by rotating the paper. Avoid straight lines when filling in color areas, and gently blend color slightly across and into the next color zone in order to keep features from being separated from the whole. You will see how easy it is for features to seem to stand away from the drawing and become isolated, even showing a pale annulus effect, unless you utilize this technique. This is in addition to the fact that many planetary features are in reality not sharply defined at all.

- At regular intervals move back and be truly objective in your assessments; you will instantly become aware of all unevenness and the other inaccuracies that do not recreate what you have just seen. Constantly try to address all these blemishes and allow the image to grow to your satisfaction from these apparent minutia; the slightest changes will frequently make all the difference. This will be a process more of gradually removing all that doesn't look like the subject (like the joke about the sculptor!), or just slowly "spot filling", more so than simply drawing the features in.

Of course I understand that drawing is now only a diminutive portion of planetary study today, and I should stress that I do indeed see the far greater value, scientifically, of CCD imaging for these purposes. Indeed, even the spectacular images of planets obtained in this way over the recent past have been achieved with exposures of hardly more than a moment of real time. This is almost what we are seeking, but not quite! None of this was experienced at the time by the observer. However in the majority of these Solar System portraits, there still remains an essential difference between the eye's outstanding resolution and that of the best CCD image using the same equipment. This becomes a different issue when it comes to deep space observing. Here the CCD image is without peer, but again, it isn't in real time. So, we are not trying to rival the wonders of CCD imaging, just to experience the maximum that we can for ourselves at the time of the observation. This is an entirely different thing.

It should also be said that to be truly expert in the field of CCD imaging, and also the incorporation of computer processing with traditional still astrophotography on film, involves complex manipulations of technology. It is an art all its own; these fields require tremendous dedication, experience and know-how, not to mention yet more additional expensive equipment on top of an already costly pursuit. These forms of observation are a separate field of work, and some excellent reference manuals exist to guide the enthusiast, should this be your persuasion.

However, for our purposes, if we extend things just a little further, there is also considerable value to the recording by real time video of what we see, as has already been mentioned. This process should not be confused with the nature of CCD imaging, as video imaging can be seen live on a monitor, and it occurs in real time. It can supplement and even help correct in unique ways any drawings we have made, and does not require volumes of knowledge to perform successfully. The results are immediate and surprisingly good. It may also prove to be the only way of convincingly imaging certain types of subjects in real time. Just such an example is my previously mentioned lack of success in drawing to any degree of satisfaction the visual effect of a rich star cluster. The video images provide far better guidance as to what you may see than anything I could draw; in fact, they actually recreate intensified viewing remarkably effectively.

If your budget allows you to try, in my view only top-rated CCD video equipment is worth the investment. (The Astrovid 2000 camera mentioned in Chapter 2 is the finest example I know of, and it features manually adjustable contrast and gain, an important refinement.) You may also be surprised by what you may see on the monitor (preferably one of high resolution), live from space! It can assist your ability to discern and make sense of detail. For recording the highest quality images, they will likely be obtained by feeding them directly to computer hard drive, retaining the first-generation image quality and maximum resolution. Outstanding results may also be obtained on videotape by using Super VHS or digital recorders. Images stored in some way are not only wonderful to have, but will also greatly serve to support what you have drawn, provide a real time record, and steer you towards even greater seeing skills. For the computer, there are other far more sophisticated video and computer imaging systems available than I have uti-lized, where simplicity has been paramount and reflects my visual observing bent. You may wish to take advantage of them. However, in my particular instance, the complete sequence of equipment for imaging deep space objects in this volume was as follows:

- Collins I_3 – eye lens component removed (with or without Barlow lens)
- Adapter (Collins) (tube to increase length)
- Eye lens (Collins/Televue) reattached to adapter
- Astrovid 2000 CCD video camera
- Control box (Astrovid)
- Collins recursive frame averager
- Sony DVMC-DA2 media converter (analog video-to-digital video fed to a computer via *Firewire*)
- Apple iMac computer
- Apple iMovie software (for still image extraction)

For the Moon and planets, the set-up was the same, except without the image intensifier and associated parts, including the recursive frame averager.

A major issue in CCD video is that of focus. For deep space it is twice as complicated as with other subjects, since the image intensifier has to be focused within itself, and then with the telescope; it is compounded by the tiny latitude you will find that you have for both of these adjustments. Of course you will be observing via monitor, and you will also find that it is appreciably harder even to find your subjects and set those variables unless you first obtain focus on a brighter object. Many deep space objects will only lend themselves to the lowest magnifications available, and even the ones that are responsive to higher powers will be easier to locate and work with if you already have them in the field of view of a regular eyepiece first. This is further made easier if you have near-exact focus points for the video set-up already pre-adjusted and approximately marked in some way on the telescope focuser. It is absolutely essential that you are precise about finally utilizing each of them, as decent results are more dependent on doing so than is possible to stress here. A digital focusing read-out (something I do not have; fine models are made by JMI) would undoubtedly assist you in the tele-scope's adjustment, but it is not essential by any means.

As for the best methods to use with video recording, time and experience will guide you best. However, I find electronic noise is kept to a minimum by keeping the gain as low as possible for any given object, and adjusting the shutter speed to a level that will show it well. Contrast can also be adjusted according to the brightness and nature of the specific object. For the Moon, the variety of illuminations suggest that there is no best recommendation for camera settings. For planetary images, and with the Astrovid 2000, I find the fastest usable settings for shutter speeds with maximum contrast will do best, along with the use of different filters, as in visual study. For deep space, the slowest shutter speeds, often with maximum contrast, will be needed in order to get the best results. You should discover for yourself the most advantageous specific settings for your own equipment; it is here that well-aligned digital or other circles really can come into their own when you have the camera attached to the telescope, although I always detach the video gear and use an eyepiece first when seeking out the next object. Switch off the image intensifier but don't unplug the video camera; the resulting electrical spark and jolt will likely make your computer freeze, making you run through the entire powering-up procedure again! Live viewing with the monitor is great for groups of people as well, but again, only when the equipment and your ease of using it is such that it creates an acceptable result. Otherwise such imaging can be quite disappointing in its inability to infect your audience with your enthusiasm.

There is no way to reproduce in any form the impact of the live view, even on the video monitor; it always reveals luminosity, detail and subtleties that cannot be adequately represented on the page by video, or even the most successfully executed drawing. An important part of the visual element simply does not transfer to the record, and how much of it does so will also depend on the specifics of any given object. While it is my hope that my own images will guide your expectations, you should not conclude that they are fully equivalent to the stunning impact the same subjects have when viewed live. Additionally, most of the deep space images were made in the typical mediocre (and sometimes poor) sky conditions of my suburban location. I didn't wait for just those few relatively optimal nights to make these illustrations, but rather took any nights that were acceptable in order to provide the most typical representations. Remember, I also do not use high resolution imaging programs or extensive processing; this serves to underscore just how much can be seen without elaborate post-viewing methods and means. It also comes closer to an honest viewing expectation, as elaborate processing can indeed reveal much more detail present in the raw image. You can probably expect to obtain equal if not superior results with somewhat better skies and much smaller apertures. Of all the additional equipment you will need, I continue to advocate the Collins I_3 Piece image intensifier; it will optimize whatever you are able to do.

In Chapter 5, the numerous lunar images, as taken with my video camera (Astrovid 2000), were included, not because the world lacks such imagery, but because it is always interesting to have a real appreciation of what can be reasonably expected to be seen with an amateur's equipment. I have not been inspired to try my hand at lunar drawings, instead choosing to capture the moment by video imaging, which produces a fine representation. The difficulties of drawing the lunar surface pose a very time-consuming task that nevertheless has some very devoted practitioners. It requires considerable skill as a draughtsman, but because the Moon has been very completely and precisely photographed, it does not seem to me to be the best use of one's time! I also feel that these lunar drawings, while often quite exquisite and admirable in the skill that they show, do not truly look like the lunar surface through the telescope. It would seem that photographs and other astro-imaging have achieved the closest equivalent of the live viewing experience, even though nothing quite rivals its near three-dimensional effect. This still does not mean you should not try drawing if you feel so inclined, but the Moon presents us with a great video opportunity, and readily produces good results.

Deep space remains the suburban astronomer's biggest challenge, but the location has no major downside for observing the Moon and planets. There is also another upside to the relatively simple approach we may take to viewing the Solar System; as we learn the art of "seeing" all that our telescopes present to us, we are better prepared for the faintest objects of deep space. Additionally, fine details, from intricate planetary features to the delicate rilles and craterlets of the Moon, not only challenge our eyes to see more but are also highly instructive in giving us first-hand experience in understanding sky conditions. This becomes increasingly important when we observe other objects, as we can only know what is realistic to expect when we understand the nature of seeing and our own particular equipment.

Chapter 5

The Moon

The Moon, being truly local in galactic terms, is our best opportunity to study another world in intimate detail (Figure 5.1). It is also a prime candidate to form a long-term relationship with from suburban locations, because it is completely accessible from what would be some of the worst conditions for almost anything else. Often flying high in the sky, it is well-placed much of the year so as not to be obscured by city buildings and other horizon blockers.

In order to enjoy spending time with our nearest neighbor in space, you will need a good, preferably photographic, lunar atlas to gain an easy comfort level during observations. Even though I have used *Atlas of the Moon* by Antonin Rükl extensively, I do not necessarily recommend this classic. This, in spite of its finely drawn viewing maps, which are probably the finest achievement known in lunar cartography. I have two primary objections: first, the maps are printed with north up, which makes referencing and reading at the telescope something of a test for patience. Secondly, the book uses a strange logic that sets out the progression across the lunar surface not in accordance with the phases. This again makes for less than total user friendliness, and easily brings about confusion. In any event, photographic atlases would seem to be a markedly better option for us. Some decent choices are available, but as of yet, I have not found the perfect one. I suggest that you make a comprehensive search of what is available in order to see which atlas seems closest to your needs.

Figure 5.1. The crater Deslandres and environs as recorded by CCD video camera.

Meantime, a classic from the 1960s is still in print, and is probably still the amateur's best guide overall: *The Hatfield Photographic Lunar Atlas*, edited by Jeremy Cook (published by Springer). The primary weakness of this near timeless work from pre-Apollo times would seem to be the photographs. While they are fine amateur astrophotographs for the time (from the 1960s) and are logically laid out, they are nevertheless inferior to the images that can be obtained with today's means. The similarly famous work and classic, *The Moon*, by H.P. Wilkins and Patrick Moore, while again a remarkable feat for its time, is not nearly as useful a guide by comparison. This is especially apparent since the line drawing cartography, so painstakingly carried out, has been completely superseded by space age imaging techniques. It is actually quite a difficult work for most people to use.

There is no denying that a lifelong and detailed study of the Moon will provide near-endless enjoyment, and its role should always feature prominently in any suburban dweller's viewing. The larger aims of this chapter, though, are not only to lead you to some magnificent lunar sights, but also to some of those smaller ones, which may reveal ways that the Moon can help expand the eye's seeing power and test our equipment. All of

Figure 5.2.
Daybreak at Aristoteles and Eudoxa.

the images were recorded in real time, directly to computer hard drive and are presented essentially unprocessed. For the most part, they were recorded in unexceptional suburban viewing conditions of air turbulence at sea level, as low an altitude as a city can have! Equipment used was my JMI NGT-18 and Astrovid 2000 CCD video camera, the same one that couples to the Collins I₃ Piece intensifier for imaging deep space subjects. As simple images, they are nevertheless quite impressive (Figure 5.2 and 5.3), compared to what used to be available only photographically.

The ease with which these images can be obtained is also quite remarkable, although of course, they do not quite compare with true CCD imaging. In overall appearance, they represent the lunar surface very well, except in the resolution of the very finest details.

Figure 5.3. Nightfall on the lunar limb and the crater Gauss.

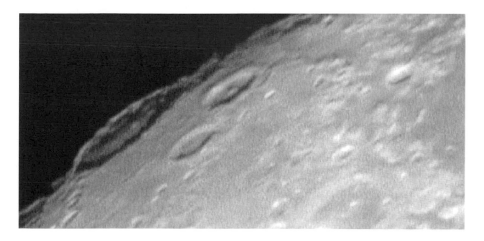

Figure 5.4. Dawn at Maginus and Moretus.

Viewed live, minutia are resolved diamond-sharp, and appear almost three-dimensional, something apparently impossible to represent in any type of recorded telescopic views. Today's best in astronomical video gear does a remarkable job nevertheless (Figures 5.4 and 5.5). The finer features discussed in my review of lunar sights may not be apparent on all of the images, especially in printed form, but at least the specific areas on the lunar surface are clearly represented for examination. It is worth pointing out that although they were taken through 18 inches of aperture, the images actually approximate resolutions of smaller apertures, owing to the loss of fine detail in the relatively simple imaging process I use. These images may be a helpful reference point in considering visual expectations with telescopes in the 4–10-inch range, and will give you a very good impression of the viewing experience through such moderate apertures in merely fair suburban skies.

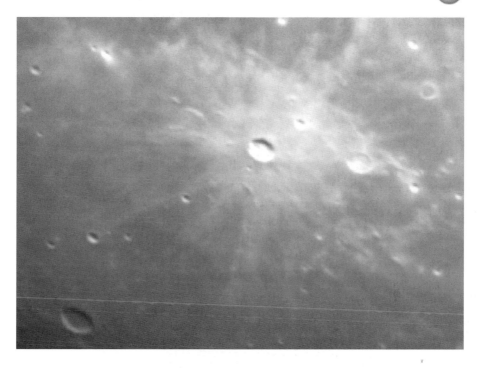

Figure 5.5. The bright ray crater Kepler.

I make no apologies for succumbing to the temptation to include many of my favorite lunar sights amongst those presented here, even though they may have been more than adequately covered in other volumes. This sampling is perhaps just a little more personal, particularly because the images were recorded with the unique illuminations of those very moments of the lunar day as I witnessed them.

Petavius (23.3°S, 60.4°E) (Figure 5.6). Long a favorite of mine, this crater has associations that hark back to my earliest days of astronomy. With its celebrated cleft, a striking feature running from the central peaks to the perimeter, Petavius was one of those lunar landmarks that I strained to discern through my tiny telescope of childhood. With a slightly less modest instrument, you will easily be able to see the vast rupture across the crater floor, splitting the surface decisively, and maybe discern other smaller clefts as well. The majority of detail will be quite clear at moderate apertures, although larger telescopes will naturally reveal significantly more.

Figure 5.6. Petavius.

Rheita Valley (42°S, 51°E) (Figure 5.7). One of a few great gashes on the lunar surface, this valley was most likely gouged out by enormous material(s) ejected during the impact that created the Mare Nectaris. Also crossed by various craters from a later date, it lies in a very rugged region and can easily be missed by the casual observer. Easily overlooked is the full extent that it travels, although this clearly shows on the video image here.

Figure 5.7. Rheita Valley.

Figure 5.8. Messier.

Messier (1.9°S, 47.6°E) (Figure 5.8). A striking feature on the Mare Fecunditatus, this is actually a triple crater lineup, with ejected material shot far out in a double ray. The dominant crater almost obliterates an earlier crater, as well as a third and similar-sized crater ahead of it, just detectable in this image. It was most likely formed by a collision with a sizeable meteor striking almost parallel to the surface.

Posidonius (31.8°N, 29.9°E) (Figure 5.9). A fine and intricate network of ridges and rilles lie on the crater floor, in addition to other rocky formations. South of the crater is a raised ridge, appearing like a crumpling of the lava plain, known as Serpentine Ridge. Its height, like many similar formations, is more akin to a large sand dune than a hill, made visible to us only at times of very oblique illumination. Occasional tiny craters can be seen at points along the spine of the ridge with good conditions and sufficient aperture.

Lacus Mortis and **Burg Rille** (45°N, 26°E) (Figure 5.10). Not far from Posedonius lies the Lacus Mortis, actually something of a ruined and lava-filled crater of vast dimensions. The rille crossing this feature is unmistakable (easily visible in the video image in Figure 5.10 as a fine line traveling straight down), but

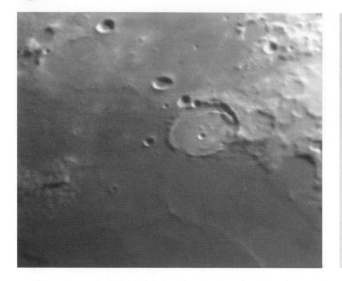

Figure 5.9.
Posidonius.

no less striking than the great fault to the east (also easily visible here), the shadow it casts being readily detectable at lunar dawn as a broad, dark feature. Look for other smaller faults northwest of the crater Burg, but most particularly the great ridge dropping off and curving west from it. Under advantageous illumination, and with steady seeing, Burg (upper right) becomes an interesting study in itself: look for its prominent central peak, but more significantly, the huge collapse inwards of its interior walls, which once experienced landslides on a massive scale. Also

Figure 5.10. Lacus Mortis (upper right) and the crater Burg.

Figure 5.11.
Theophilus.

significant is the crater Aristoteles to the west, a structure not unlike Tycho, but without surrounding rays or a significant central peak.

Theophilus (11.4S, 26.4°E) (Figure 5.11). This is one of the most prominent features on the Moon, with towering central peaks and a complex rampart structure. It keeps company with two other striking formations, the older and less well-preserved craters Cyrillus and Catharina. Together, around first quarter, they are unmistakable landmarks.

Plinius (15.4°N, 28.2°E) (Figure 5.12). Although not traditionally one of the more famous lunar locations, Plinius is nevertheless a visually pleasing, sharply defined crater with central peak, terraced walls and interior detail, located in between the Mare Serenatis and the Sea of Tranquility. Perhaps my main interest, though, in this otherwise unlikely-to-be-favored structure betrays a little of my professional background: a rock formation to the east amid other interesting crumpled and pitted landscapes, and adjacent to the

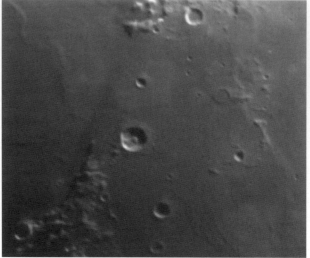

Figure 5.12. Plinius.

small crater, Carrel, looks for all the world to me a lot like a very familiar musical notation – an eighth note! (I don't believe any musicians were there before the Apollo astronauts landed!) Look for it mid-frame toward the right side of this video image.

Hyginus Rille (8°N, 6.3°E) and the **Treisnecker Rille** system (5°N, 5°E) (Figure 5.13). Hyginus Rille presents a most interesting geological riddle, since for much of its length it comprises a chain of rimless craterlets, interrupted mid-length by the crater, Hyginus. The finest of these craterlets become more difficult to resolve, and aperture, steady seeing as well as good optical quality will pay dividends. So how did this unlikely chain of craters just happen to align into what we see? Far from being a freak of meteoric bombardment, it appears instead that the rille is probably just a collapsed lava tube, formed by systematic collapse along the length of the underground structure. To the northeast lies the wide Ariadaeus Rille, an easy sight for small telescopes, with one end visible here in the image extending into the top right corner. The spectacular and complex Triesnecker Rille system may be found immediately to the southwest of Hyginus, and east of the crater after which it is named. Viewing these intersecting and finely formed rilles immediately after dawn or just before sunset is the most significant factor in being able to resolve them successfully. They are

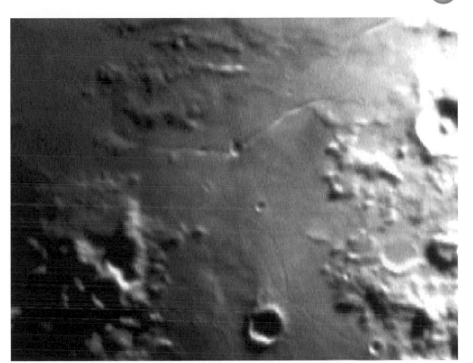

Figure 5.13.
Hyginus Rille and the Treisnecker Rille system.

truly spectacular and mazelike through my own telescope with only moderately good suburban viewing conditions. As probably the most celebrated rilles on the Moon, and frequently imaged by observatories since the beginning of lunar photography, even users of small telescopes will have success in revealing many of them.

Mount Hadley; Hadley Rille (27°N, 4°E) (Figure 5.14). The landing site of Apollo 15. This is one of my favorite lunar mountainous sites. Although not prominent on this video frame, Hadley Rille is not too difficult an observing test with moderate apertures, given reasonable conditions. As with other Apollo landing sights, with sufficient resolution and magnification it is possible to gain a real insight into lunar terrain, when comparing the view with photographs from the mission. For smaller telescopes, though, Hadley Rille becomes harder to resolve and trace to its full length, as it winds around the flat terrain and crater Hadley C. Look also for the prominent Bradley Rille nearby, (23°N,2°W) as well as other similar formations.

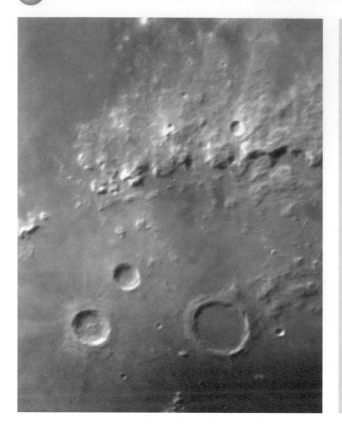

Figure 5.14. Hadley Rille.

Hadley Rille is likely to be in the same field of view as Archimedes (29.7°N, 4°W), Aristillus (33.9°, 1.2°E) and Cassini (40.2°, 4.6°E). These three craters make a varied group. Archimedes is a prominent lava-flooded crater with high walls, not unlike Plato in appearance, but smaller and with a paler floor color; it also has a number of small craterlets that can be challenging to resolve, but is in no way the equal of the nearby and better known Plato for craterlet counting and spurious claims. Aristillus has several low peaks in its center and is one of the well-known "ray craters", becoming more prominent with increasing lunar phases.

Alpine Valley (49°N, 3°E) (Figure 5.15a). Just east of the walled crater Plato, this most famous of lunar valleys looks like a huge gash right through the Lunar Alps. Actually, it was probably formed like many other smaller and similar valleys, as a flow of lava, and not the sort of collision its appearance tends to suggest.

Figure 5.15.
a Alpine Valley and
b the rille.

Obvious with the slightest optical aid, the valley is striking during second quarter. Larger apertures combined with good seeing, even from the city, and optimum lunar morning or evening illumination, will reveal a fine rille on the seemingly level lava-flooded floor, running most of the length of the valley. (Just detectable in Figure 5.15b as a fine bright thread, this feature is readily visible in 18 inches; it will probably also show in some lesser apertures.) Low left of center lies the ruined crater formation Cassini.

Ptolemaeus (9.2S, 1.8°W) (Figure 5.16). No chapter of the Moon would be complete without drawing attention to this grand crater. Long a favorite target of amateur astronomers, it has regularly been the focus of reported "volcanic" activity, although such claims have never been substantiated. Its floor, as with many crater basins, was long ago leveled out into a plain by lava

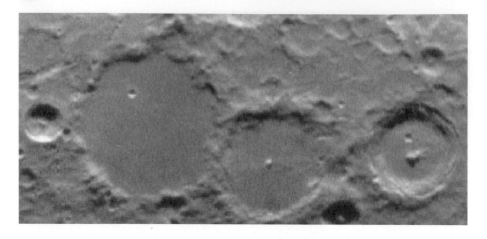

flow, and features many small craterlets and almost-obliterated earlier crater formations. Nearby are the no less interesting craters Alphonsus and Arzachel (with its marvelous central peak), both also part of the volcanic myth. These three craters become prominent just after first quarter.

Figure 5.16.
Ptolemaeus.

Straight Wall, Birt Rille and **Pitatus** (21°S, 9°W) (Figure 5.17). The Straight Wall remains one of the most celebrated features on the Moon for its great length and height. Approximately 60 miles long and presenting a face around 800 feet high, it must be an awesome sight from ground level; nothing close to these dimensions exists on Earth. Immediately west of this grand fault lies a test of vision and equipment, Birt Rille, quite a fine feature (barely visible here). However, an even finer test is the small craterlet at each end of it, which moderate apertures should reveal when our suburban atmosphere allows. Also look for shading variations within crater Birt; these will be visible as subtle irregularities, and are not usually commented upon. I have seen them clearly. Pitatus, to the east, is striking for its fine rilles and complex ramparts.

Plato (51.6°N, 9.3°W) (Figure 5.18a). This vast and famous crater is another example of one which has been partially filled in with a dark lava flow from a later time, forming a smoothly textured floor. A favorite target for amateur astronomers, there are at least five reasonably prominent craterlets to look for on this

Figure 5.17. Straight Wall, Birt Rille and Pitatus.

wide plain. Many more have been observed and counted by multitudes of Moon watchers over many generations; because of the exact number and location of the craterlets, Plato was in the past amongst the amateur observers' most highly watched and controversial lunar sights. Much of the controversy over these craterlets has been settled by lunar orbiters. The formation itself is so striking in appearance, regardless of the phase of illumination, that it stands out as a dominant feature on the lunar surface; small craters can be seen as bright dots even at full moon. Do not miss nearby Mts. Piton and Pico (Figure 5.18b), standing in splendid isolation on Mare Imbrium.

Tycho (43.3°S, 11.2°W) (Figure 5.19). One of the foremost lunar ray craters, the impact that formed Tycho spewed out material over many hundreds of miles, further than any other crater of its type. This type of crater grows in prominence towards full moon, none

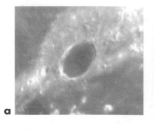

a

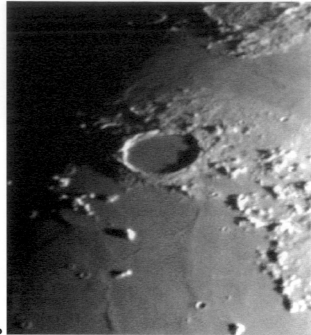

b

Figure 5.18. a Plato and **b** its surroundings: Mt. Pico immediately below; Piton bottom right.

Figure 5.19. Tycho.

more so than Tycho; in this instance an otherwise only moderately imposing crater becomes very bright and its extending rays seem to dominate the entire lunar landscape.

Clavius (58.4°S, 14.4°W) (Figure 5.20a). This massive, grand and prominently placed crater must be familiar to anyone who has spent any time at all observing the Moon. It is known to many others as the site of the fictional lunar base in Arthur C. Clark's *2001, A Space Odyssey*. it reveals a multitude of details, the best-known being a chain of ever-diminishing inner craters,

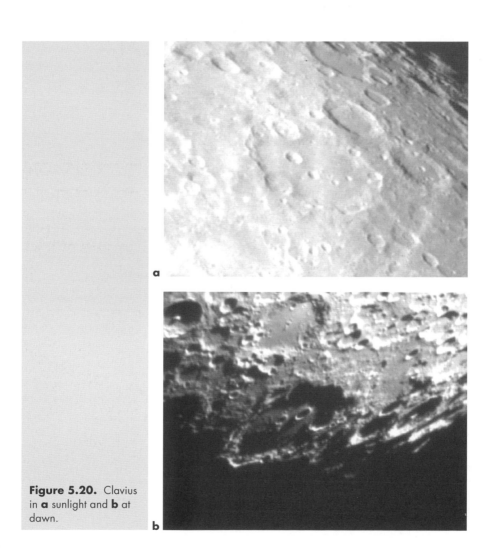

a

b

Figure 5.20. Clavius in **a** sunlight and **b** at dawn.

Figure 5.21. a Morning light on Eratosthenes. Note the chains of craterlets below – these are ejecta pits from the formation of nearby Copernicus (just appearing at dawn, bottom center); also note the large lava-filled crater Stadius, center right. **b** Eratosthenes just 24 hours later.

evenly spread across its floor in an arc. Clavius is one of the largest crater formations on the entire lunar surface, though not the largest, as some comment. Its dominance is striking even though its walls lie almost entirely below the surrounding rugged lunar plain. You will probably appreciate observing it under many lunar illuminations (Figure 5.20b), not only for its spectacular nature, but also because it lies in some of the most rugged and interesting terrain on the lunar surface.

Eratosthenes (14.5°N, 11.3W) (Figure 5.21). This striking crater lies in the vicinity of the great ray crater Copernicus, but precedes it in the arrival of dawn. The video image in Figure 5.21a was taken shortly after daybreak, and at the lower center can be seen the first sign of light on the eastern edge of Copernicus itself. Eratosthenes is no less notable for terraced walls and central peaks than its grander and better-known close neighbor; it also forms a kind of marker for the western extreme of the Lunar Apennines. Like some other striking formations, it is also significant for its great range of appearance under different illuminations, and it graduates rapidly from a highly striking formation to one that is hardly visible in only a very few days.

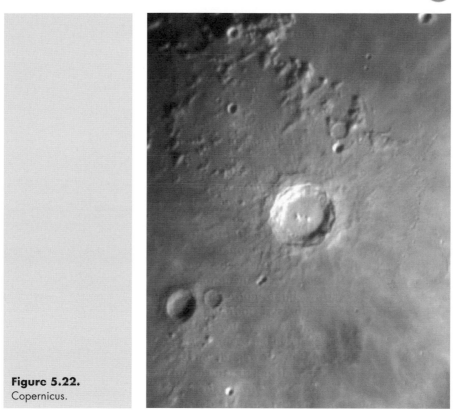

Figure 5.22.
Copernicus.

Copernicus (9.7°N, 20°W) (Figure 5.22). One of the most spectacular of all lunar formations, not only because of its dramatic structure and visual placement, but also because it is one of the Moon's most prominent ray craters. Copernicus' massive walls rise over 3,000 feet above the surrounding terrain, with much evidence of landslides on their interior. The floor lies well below the lunar plain by as much as 7,000 feet, and a series of central peaks rise from it to a height of nearly 4,000 feet. Surrounding the crater lies the inflated lunar crust, as well as extensive radiating debris and craterlet pits from what must have been the cataclysmic impact long ago which formed it. It must have pushed upwards enormous amounts of material from deep below the surface, which finally rained down on the surrounding plain.

Gassendi (17.5°S, 39.9°W) (Figure 5.23). This ruined crater is quite a magnificent formation, so much lava

Figure 5.23.
Gassendi.

having filled its interior that it produced a circled plain, riddled with clefts and mountain peaks. On the north side, a more recent collision created a smaller crater which pushed the wall in and part of the floor upwards. Be sure to spend time with nearby Mare Humorum. Its extensive ridges, as well as rilles such as Doppelmayer and Hippalus rewarding the time you will have spent.

Schröter Valley (26N, 51°W) (Figure 5.24). A curious formation, Schröter Valley takes on the appearance of a cobra, as it widens and culminates in the crater Herodotus, the cobra's head. Depending on the lighting, the chiseled landscape to the north can give the illusion of a twin valley in the opposite direction. The visual test is to see how far you can trace out the real valley, which narrows along its length towards the cobra's tail. East of the valley is the extraordinary ray crater Aristarchus; its sloping sides, bright interior, and central "bump" make it a striking formation.

Figure 5.24.
Schröter Valley.

Also in the vicinity of the valley (just southwest), and likely to be in the same field of view, is a typical, but sometimes frustrating lunar rille:

Marius Rille (17°N, 49°W). This rille is one of the better examples for testing visual acumen as well as equipment. Thin and winding, it will demand keen viewing skills and equipment of at least moderate aperture. As a much quoted test, the challenge, as with others, is to see its entire length. The only chance to do so will be shortly after lunar sunrise or sunset, when contrast will be at its best. (It is not difficult with 18-inch aperture.)

Schickard (44.4°S, 54.6°W) (Figure 5.25). A vast, partially lava-flooded crater with walls and floor pockmarked by small craterlets. After first quarter this is a mighty feature low on the south limb. It is located in one of the most rugged and highly cratered areas of the Moon, and will be striking to view under many vari-

eties of illumination. The view here was taken at dawn on the region.

Figure 5.25. Schickard.

Many larger areas of lunar landscape will provide constant visual engagement for the amateur observer; these are not necessarily specific features, but regions that by their very natures demand attention. Shortly after dawn the region adjacent to the Mare Frigoris (62°N, 0°) (Figure 5.26) is one such place. In my view,

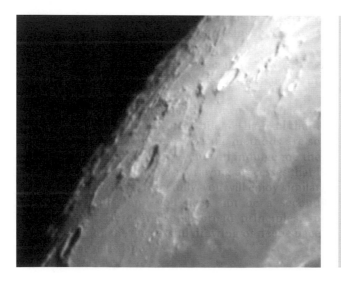

Figure 5.26. Mare Frigoris.

Figure 5.27.
a Promontorium Laplace and **b** Harpalus.

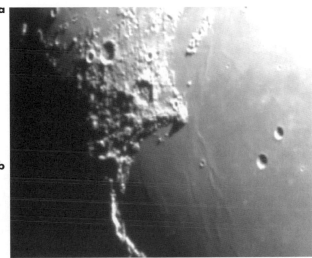

there is little like this barren region anywhere. Stark and other-worldly, with its deep shadows, the area is punctuated with bright craters and rays.

Another wonderful scene is nearby Sinus Iridum. On the eastern extremity of the surrounding "wall" is a tall feature, Promontorium Laplace. After dawn it casts a large, pointed shadow over the corner of the plain (Figure 5.27a). Also look for the striking crater Harpalus, to the west (Figure 5.27b), which seems to show a double wall and flat floor, like a woven basket in appearance; (sunrise has not yet taken place on it in Figure 5.27a, right; see instead Figure 5.27b). To the east (top center, Figure 5.27a) is an unusual and almost straight isolated mountain range, the Recti Mountains (45°N, 34°W). This feature lies at the western edge of Mare Imbrium, a vast lava plain with isolated jutting peaks and expansive creases and crumplings, south of Plato.

Ghost craters: in the region of another striking walled ray crater, Bullialdus (on the left in Figure 5.28), is a varied array of mostly lava-filled remnants of craters. These old structural remains, almost completely hidden now, were formed and almost obliterated before the creation of the region's more complete crater structures. Similar formations can be found across the lunar surface, but this remains one of the areas of greatest concentration (20.7°S, 22.2°W).

There are, of course, almost limitless other sights on the Moon that could be included here, and many more tests for the eye and telescope, but since this is not

Figure 5.28. Ghost craters near Bullialdus.

primarily a textbook or an atlas on the Moon, it was necessary to call a halt somewhere. As a complete study in itself, lunar observing fills volumes, so I strongly recommend that you avail yourself of at least one major book devoted exclusively to it. It is the only celestial object which reveals itself to us so readily and so fully. Perhaps its strongest card for us in the confines of the city is that it is completely unaffected by city light and air pollution. For suburban or urban viewing it can be a salvation; since it can be counted on as a regular sight in the sky there are still those who scarcely observe anything else. A downside is that when sky conditions are perfect for deep space observing, there is no guarantee that the Moon won't be high in the sky, and even full phase, to wipe everything else out – the ultimate light pollutant!

Figure 5.29. Views like this, taken just before first quarter, almost give us a sense of being in orbit!

Chapter 6

The Planets

There are few objects more accessible to our telescopes than the planets; they are ideal targets for us as they need no special provisions for the suburban viewer, except the steadiest air. However, only a minority of them qualify as the grand real time sights I had in mind for the purposes of this book. Those that do are visually stunning, and should feature large in any city-bound observer's viewing.

First out from the Sun is Mercury, a small planet, which follows the Sun so closely that it is placed favorably only infrequently. Even at these times, it is an object that yields detail sparingly. One can expect only to see a few markings on relatively few occasions. That doesn't mean one shouldn't try, of course, since there are observers who have endeavored to make a major study of the "Winged Messenger". However, in my opinion, the results still do not justify inclusion in this writing. Outwards from Mercury orbits Venus. For such a dazzling naked eye sight, it is one of the most profoundly disappointing sights in the telescope! Some observers take great delight in viewing the successive phases and dramatic size variations of the telescopic view of Venus, but these are scarcely more interesting to me than seeing our own Moon's phases with the naked eye. There have been some remarkable studies made regarding Venus' rotation beneath the cloud cover (from such vague markings as can occasionally be seen). However, just as with Mercury, the images available from spacecraft missions certainly diminish the importance of anything that can be seen from Earth, and again point to the difficulties we face as terrestrial observers regarding many of our nearest neighbors in space. Neither of these sights is significantly improved in better surroundings than ours!

Only when we turn to Mars do we experience something of a viewing revelation. However, views of it from Earth orbit, and the even more astounding results obtained by various spacecraft "on location", tell us something of the futility most of us have in trying to accomplish anything much scientifically; there isn't much that hasn't already being done far better and more completely. Even the one remaining area that amateurs still claim (daily observation of atmospheric changes and phenomena) is rapidly being made redundant by more continuously vigilant and powerfully equipped spacecraft parked in Mars orbit. However, some amateurs are still contributing relevant research, though strictly of a limited nature these days, by monitoring these conditions on the planet. If this is your "bent", contact one of the planetary observing groups listed in the Appendix B, or elsewhere. There may still be something you can contribute.

Today, amateur astronomers, using CCD imaging and modest apertures, are capturing detail and resolution thought to be impossible even by the world's largest telescopes of the most recent past. Often they succeed astoundingly in brightly lit suburban areas, on balconies, rooftops and the like, which in the past would have been considered totally unsuitable, even for live viewing. Visually though, there is nothing quite like the experience of actually observing a planet such as Mars. It remains one of the most exciting celestial sights directly accessible from our city locations. When conditions are favorable, the variety and wealth of detail discernable may astound you, assuming you have reasonably good equipment. With a little persistence, the "God of War" will provide some spectacular viewing. As with all suburban planetary viewing, just keep bright lights shielded from your line of sight.

Still further out into the Solar System, arguably the most readily accessible planet for live-viewing purposes is Jupiter. There is no other planet, granting us such frequency of apparition, that presents us with its disk size and variety of well-defined detail. Additionally, it is always undergoing change, and its atmospheric blanket shows us a very violent place in a state of never-ending turmoil. Much more accessible than Mars, it provides an endless array of changing features. B.M. Peek (author of *The Planet Jupiter*, published by Faber & Faber), made a lifelong study of it, to the exclusion of virtually everything else. His book, though scientifically dated and difficult to obtain, remains the definitive textbook for amateur observers.

Saturn is also a superior object for planetary study. Most novices are more enthralled by their first views of it than by anything else. However, it has to be said that compared to Jupiter and Mars, it is somewhat static in appearance, with cloud-belt changes occurring rather slowly, such detail being substantially fainter and less complex. Nevertheless, there is much to see. On the rings, look for any visible shadings, divisions, or shadows they cast on the planet, as well as color subtleties. While the cloud-belts on the disk itself exhibit similarities to those on Jupiter, the relative faintness and seemingly quiet order of things set them apart. Of course, spacecraft have revealed great drama on the surface, but we still cannot observe this successfully from Earth. The rings themselves provide some of the best viewing, since their angle relative to Earth is always changing, and shadows from the planet are variously cast on them, depending on the time during apparitions. I usually find that just one really good drawing of Saturn during each apparition is sufficient to show any major changes that have taken place. It is also notoriously difficult to draw, so it may actually be something of a blessing to us that it does appear so relatively static!

The distant outer planets are a different story when compared to the last three, however. Uranus and Neptune certainly are worth the trouble to seek out, and their bluish-green disks always remind us that we are looking on other worlds, though they are both sufficiently remote as to not present us with much more than just tiny disks. Uranus may possibly, however, show some faint banding with sufficient aperture and power. Pluto remains a unique challenge for many observers. Just to see it at all, merely a star-like point, provides a thrill and sense of accomplishment to many observers. It can usually be sighted by referencing monthly astronomical periodicals, which may also provide locating tips, and is easiest to find when it is reported to be in the vicinity of a relatively bright star. Viewing it can be enhanced by the use of an image intensifier; in the suburbs this will make actually seeing it more likely. Asteroids can provide similar pleasures and challenges, as well as the occultations of stars caused by their orbits. Again, this writing is not aimed at such observations, but if this appeals to you it would certainly be possible to pursue it effectively from suburbia with the use of an image intensifier or even other means. In the suburbs, you should treat viewing the subjects of this paragraph more as you would deep space objects, and you will need to be well shielded from direct bright lights. You will not only need the greater conditions of darkness, but remember that image intensifiers do not react well to direct lights.

In this chapter I have included a few real time video images of the planets, as taken by the author. Cameras such as mine record only in monochrome, and as such are not especially helpful in revealing the subtle color appearances of the planets, which are present in delicate, though unmistakable, hues. Although it is possible to use color filters to make three separate images, combining them to produce amazing color images, these still will not likely reveal the subtleties to which I refer. Nevertheless, sometimes these processed video images are nothing short of remarkable. This, though, is entering a whole new area beyond the direction of this book. Color CCD video cameras can produce beautiful, well-scaled planetary images readily, but they are still unable to capture the most delicate color subtleties, as well as some of the fine detail that only our eyes seem able to perceive. (Remember, because their total light and spectral sensitivity is different to monochrome versions, color video cameras are not suited to deep space. My monochrome Astrovid video camera also doubles in its far more important role, for my purposes at least, of real time deep space imaging, when combined with the I_3 image intensifier.)

However, such video planetary views as are presented here are real time views, taken from the video footage, and in the spirit of this book, recorded in routine suburban viewing conditions. Examining this footage frame by frame can reveal surprising detail quite readily. The best results will naturally be obtained by eliminating recording generations and therefore utilizing the maximum resolution. Those obtained by interfacing directly to the final medium will be the best. (As a starting point, check with Adirondack, for information concerning this.) In fact, results of video can be incredibly exciting, as well as serving as an excellent reference when drawing any subject, particularly in the placement of planetary features on a blank planetary disk. Sometimes you will also notice things you didn't see live through the eyepiece; whether this is because of differences of sensitivity between the eye and the camera, or just the comfort of perusing images with both eyes in a relaxed environment, is not always clear to me. But the video is still a very different experience, and has an entirely different visual impact from the view through the eyepiece. I realize there is a fine line between this form of imaging and leaving the real time observing experience behind altogether. When dealing with the quality of image obtained with today's astronomical video or other CCD gear it isn't very difficult to understand how the true CCD imaging enthusiast is born. Although CCD images do provide the most accurate recorded images available today from Earth, they sometimes present surprising differences to actual visual observing. Often, they simply appear too clear, yet not as "clean" as the eye's view, whereas the actual live experience is something much more subtle and vague, in spite of an impression of exquisite focus at times. I don't wish to discourage you from becoming a CCD enthusiast, but you need to recognize where your main interest lies. Mine has never strayed far from the live event, the record still crystallized for me by drawing.

In the planetary drawings presented here, I have tried to capture the fragile, subtle essence of the live moment. Because of the effort involved, generally I don't draw unless seeing is reasonably good. I have not attempted to make a full and comprehensive background survey of the planets that I do detail, as again, the primary aim is to provide something that prepares you for the views you may be able to experience, live in your suburban backyard, and not to repeat available materials. I also do not pretend to have made precisely geographically exact renderings, but nevertheless they are close to this aim, and certainly are faithful to the "look" that was present when I drew them. Additionally, I want to show you what else can be undertaken aside from full-disk drawings. In recording planetary images, official protocol requires that you note such things as the location, date and time, viewing conditions, telescope, magnification, filters used, longitude of the central meridian, as well as other pertinent information for specific modes of recording (i.e. CCD imaging). Some of this may be overkill, depending on the use you have planned for the images. I take a more selective view about the information, mostly a means of cross-referencing one image against another; there-

fore I only record the information helpful to me. You
will often read of the importance of converting your
local time into Universal Time (UT) or Greenwich
Mean Time (GMT). Unless you have an official purpose
in mind, this is not necessarily useful to you, and may
actually be a nuisance for your own cross-referencing
and recall. You can always make these conversions, or
compute longitude readings at a later date, if you need
or wish to.

Let us move on to specifics for viewing each of the
three planets most readily accessible to amateurs:
Mars, Jupiter and Saturn. Observational approaches to
each will be presented in separate sections. The aperture and type of telescope you use will be significant in
how well you see the same features as I show, as well as
your success in seeing the subtle range of color that I
have drawn. However, a substantial proportion of the
visible surface markings is generally large enough to
show in telescopes of relatively modest apertures. The
larger the aperture though, the greater will be the color
range, fineness of resolution of the detail, and brightness of the image. Unless otherwise indicated, the planetary images I include in this chapter were all observed
through my 18-inch reflector. If this is a much larger
telescope than that available to you, do not be deterred;
on really good nights, amazing things are possible with
much less.

The drawings include the date of the observation,
the time (in Pacific Standard Time, my own local time),
the magnification used, and the seeing conditions
(using Herschel's scale from I–V, I being best).

Mars

The simple, raw CCD video images of Mars in
Figure 6.1, from the 2001 apparition, were recorded on
a simple VHS video recorder, and are included here to
show that some remarkable detail can still be realized
even after the loss of resolution caused by this rudi-
mentary imaging process. The video footage, seen as a
visual average and yet more revealing than the still
frames, was advanced frame by frame, and those with
the maximum detail were selected. For the best possible results, you might wish to record your images
direct to computer hard drive, along with some higher-
end software than I use.

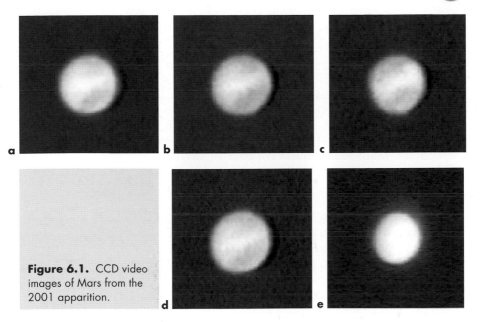

Figure 6.1. CCD video images of Mars from the 2001 apparition.

In addition to the Mare Acidalium region and Margaritifer Sinus, the Southern Polar cap, the Solis Lacus and environs are also just possible to discern on some of these images. The two lower images show the increasing effects of the classic global Martian dust storm, which ultimately obliterated most surface detail by mid-apparition. They also show the increasing phase effect as the planet moved further from Earth.

Something of a mystical aura still lingers over this planet, even though perhaps we know more about it now than any other planet except Earth. Gone are the days of Martians and their cites, canals and their never-ending quest for water. Gone even are dreams of finding evidence of long-gone civilizations. The first spacecraft dispelled those ideas years ago, but they gave us no inkling of the huge variety of terrain that successive spacecraft would reveal instead. I cannot deny that when looking at this world, something of a longing for the old myths and visions of Percival Lowell still remains. It is hard to look at the albedo features and not feel some of the same pull on the imagination that fired up all of the legendary Mars observers of the past. Some of the latest theories and discoveries have once again added new interest to the search for life, or past life (microscopic organisms, no matter how primitive), and the possibilities for converting the surface to a suitable world for human habitation some time in the future (terraforming).

As Earthbound observers, we don't have too many opportunities to observe Mars, and even fewer occasions to observe it well. Being in Earth's neighborhood infrequently (every 26 months), and only for short periods, we must take advantage of any opportunities we have to study it. A compounding factor is that only some of those "visits" are relatively close to us, and even at its closest, Mars also presents us with a rather smallish disk, smaller than Jupiter is at its most distant. Additionally, the best oppositions tend to place it at lower latitudes than desirable for Northern hemisphere observers, and sometimes are accompanied by dust storms on its surface severe enough to cover up all of its features! All in all, a difficult object.

In spite of everything, we have been able to discern remarkable amounts of detail and information from our Earthbound observing sites. The Red Planet does us the considerable favor of turning its face a little differently to us with each opposition, and it is from these variations that amateur astronomers have been able to have, first hand, a fuller appreciation of its features than they might otherwise. With Mars, there is no escaping the fact that aperture does count, although it is possible to see tantalizing detail with apertures as small as 3 inches. And wonderful detail it is! It presents no question as to what we can see, as the main features are quite apparent to an experienced eye at first glance. Mars, being so bright at opposition, is one of those sights which allows the use of high magnifications – more so than one might expect. Always try using somewhat lower powers initially, slowly increasing it as your eye discerns more detail and contrast. The final power may be surprisingly high. Mars will easily stand 458× with my telescope on most decent nights, although instruments of substantially smaller apertures often will also take powers equal to this. Frequently, even more power works readily (up to 680×), though it is seldom called for, or even useful.

In his excellent book, *Patrick Moore on Mars* (published by Cassell), Moore provides some templates for drawing the full disk of Mars and its various phases, which he volunteers freely for copying. (I am not sure of copyright issues on this apparent open invitation, however.) There are other ways to proceed, of course. Planetary drawings tend to look their best when the disk is presented against a black background. It is easy enough to cut out a white disk, glue it to black blanks and photocopy them. The various appropriate phases can be drawn in and filled with black ink or pencil, and also photocopied. Patrick Moore uses a basic disk size of 2 inches, which personally, I favor. The Association of Lunar and Planetary Observers uses 42 mm. Really, it doesn't matter which size you choose, as long as you are not supplying your drawings to an organization which specifies dimensions. I draw the planets because I enjoy recalling the essence of what I saw for my own records, and the process itself. It is impossible not to learn from it.

Once I have made enough blank disk sheets, usually containing four blanks of about 4 inches square, I cut these out and store them for future use. Once finished, I simply paste the drawings in my record-keeping book, along with seeing conditions, magnification(s), date and time, plus any filters used if these were a major factor in the viewing.

When the Martian disk is full, it is then at its most satisfying to view during the apparition, since it is closest to the Earth and most readily reveals detail. I find it best, as with other planetary drawings, to first spend some time allowing the eye to adjust and settle on what it is seeing. Do not start drawing too soon! The rotation of Mars is not nearly such an immediate factor as it is with Jupiter, but you will nevertheless notice the features slowly drifting across the disk. This slower rotation will, however, limit the number of occasions you can see the same face of the planet, usually amounting to only a few complete rotations per apparition, since each night the planet will appear to have turned only a fraction. This is because the Martian day is almost the same as Earth's, except it is slightly longer. Once you are ready to draw, in order to capture as accurately as possible what you see, first make a quick sketch at the eyepiece of the positions of the features. Only after these features are established should you spend time drawing the finer details, and indicating subtleties of color and shading. Curiously, the colors will probably seem to be more or less intense from session to session. Since we can be sure it isn't the planet that has changed with such rapidity(!), these differences can only be explained by noting variations in viewing conditions, or those even within our own perceptions. I always feel that the steadiest nights bring out the most vivid colors, but I am sure there are other factors involved. The colors themselves are also being seen through the thick, light absorbing atmosphere of

Earth. Because of this, and also the illusion of simultaneous contrast, the colors of any planet should not be taken at face value. However, I take the approach that what I am trying to draw is what I see at the time, regardless of anything else.

To make the surface subtleties a little more obvious to you, red or orange filters will help bring out the darker markings. Blue, violet and green filters will aid in seeing atmospheric phenomena and clouds, although I have sometimes seen dark Martian surface features particularly well with a blue filter – not the way it is supposed to be! I have even used a Moon filter to knock out the glare, sometimes to great effect, since on a really steady and transparent night this glare can mask what we are trying to see.

As with all color drawing, beware of too much of it! Part of the authenticity and beauty of any planetary drawing is the subtlety that this time-proven method of imaging can bring to the features and shading; in my view this aspect, at least, is still superior to CCD imaging. (When CCD images are processed to show very subtle color, something of the crispness of the image seems to be missing, even though they are not necessarily blurred.) Don't waste what is unique to our live viewing by drawing the details so blatantly that all suggestion of the reality of it is lost. If you do add too much color, you can use your pencils' erasers in very gentle motions to lighten it without spoiling what you have. I have found that my most successful planetary drawings generally represent color very sparingly. Anything else just doesn't look real.

For Mars, the following drawing techniques are usually successful:

- Lightly sketch the principal albedo features in pale gray. Be careful not to represent them as larger than they actually are; this trap is easier to fall into than it sounds. Lightly fill in these areas. Gently add a hint of brown and green according to the way the features appear, all during this process adding more gray generally. Blend these with your finger. Use gentle strokes of the erasers to lighten and shape any part of these features that require it, until what you see begins to approximate the effect of what you have actually seen through the telescope. Experience shows that most of the darker features contain mostly gray; colors are there as well, but the slightest amounts suggest them strongly.

- When you get close to something that pleases you, begin to add a light peach color to the open remaining areas. You will see that the ruddy surface color varies subtly across the planet, and discreet additions of true red and orange will provide the range you need. The ruddiest color will usually be perceived toward the center of the planet, and in larger apertures it is a striking hue to be sure. Around the edges of the disk, you will notice the colors become ever paler, and typically "limb arcs" (due to atmospheric haze) are commonly seen; concentrations of dust and thin cloud are compounded at the extremes, and have the effect of whitening the limbs.

- Where the paper takes color unevenly, try dotting the required color into that specific area and then blend with a fingertip. You will find that sharp pencil points can be helpful for this, although they can be detrimental to other discussed techniques. You may find it necessary to alternately sharpen and flatten the pencil lead, to a degree which rapidly reduces the life of the pencil. Certain features, such as the Syrtis Major, will at times need additional colors, such as medium blue (for the Blue Syrtis Cloud phenomenon), or even black to deepen the hue sufficiently. Throughout all of this, I stress again that less color is more, and some indistinction between the dark features and the ruddy base color of the planet will create a far more realistic result. Only as you become confident your drawing is near to completion is it time to deepen the colors to the final shade. It is always better to add than to subtract, and always best to be certain of placement and proportion of the features before reaching this last stage.

- I don't personally favor the practice of indicating bright cloud regions with dotted lines on their borders. Although useful in cases where these drawings are to be submitted for research, they substantially detract from our efforts to capture the live appearance of objects in space. Use your best judgment.

I have always found Mars an inordinately difficult subject to "read", and representing the surface features accurately proportioned and shaped is harder than it seems it should be. (A video record can help in this regard.) The features seem to stand out clearly enough, but actually transferring them correctly to paper is one of the most challenging things there is amongst the subjects that we have. Compared to CCD images, my own drawings reproduced in this volume may also be guilty of certain inaccuracies. However, I try to take whatever measures I can so as not to be guiltier than necessary! The "seeing" of non-existent features, and feature-shaping inaccuracies are also present in most drawings by other observers. In the case of Mars, all kinds of features have been "seen" that were never there, the most famous (or infamous) being the so-called canals, of course. Before the advent of spacecraft they were drawn frequently, with only one or two real surface features ever forming any basis and justification for any of them. The rest, of course, had no basis in reality, although their advocates firmly believed they had seen them. I suppose it is not surprising that the controversy arose, given the difficulties of observing Mars, even though I have never seen a trace of anything that could be considered canal-like. I have also never been even remotely led into thinking that I could see anything that took on the appearance of the geometric shapes and patterns so often drawn as representations of the surface features. Always remember as well that in any drawing we are not trying to be Picasso; there is absolutely no room for artistic license, only for good draughtsmanship! But even today, many drawings are made that seem to show things that simply are not there, despite the sincere belief that their creators saw them. So I am always proud when any drawing I have made "stacks up" well against photo-accurate CCD portraits made at the same time. You should be, too.

You will find, with consistent observation, that very real changes do take place on the planet. These have been observed and confirmed for many years, and many are seasonal. The exact causes of these changes are probably many and varied; with experience you may be able to discern some of them for yourself, although we now know that they certainly are not due to vegetation. Any good reference source will show you where to look and when, but the unexpected can still happen! However, some of the changes you may believe you see, such as variations of color, more likely reflect the seeing conditions of the time than anything else you might suspect, or even the altitude of the planet in the sky.

As with almost everything, the Martian drawings in this volume were made from observations through my own 18-inch reflector, and from my home location in Southern California, noted in local time (Pacific Standard Time). They are presented to you un-retouched at later dates, as I believed I had seen them at the time. Mars remains one of the prime jewels for the real time suburban amateur observer, together with the would-be planetary portrait-maker. From drawing to drawing I have spent more time trying to capture the essence of this little world than any other single object I can reach through my telescope. Because Mars' subtle nature always demands our careful gaze, even at the most favorable oppositions, be patient and steadfast; slowly but surely it will reveal its face from what may at first glance appear to be a blank stare. Despite the problems of viewing and drawing the red planet, any worthy images we can glean may also provide us with some of the greatest satisfaction from the results of our labor.

The four examples from the 2001 apparition shown in Figure 6.2 provide a good representation of one entire rotation of Mars.

Variations in Aspect at Different Oppositions

At each opposition, because of mutual axial tilts, Mars is placed in its orbit a little differently relative to Earth, which changes the angle that we view it from in space. Because of this varying slant, the northern or southern hemispheres take on greater or lesser emphasis, depending on the apparition, revealing constant variety in the face of the planet. The four drawings in Figure 6.3, from two recent and separate oppositions, and of much the same regions (in longitude) of Mars, show to what degree different planetary approaches affect that angle, and how the features are consequently presented to us. When these northerly or southerly features are positioned nearer the center of the visible disk, not only do they show truer proportions, but they reveal themselves with greater contrast and clarity. You can expect a far greater range even than you see here over an entire cycle of oppositions.

The four drawings in Figure 6.3 were taken from oppositions only two years apart; they show the extent that this effect is apparent.

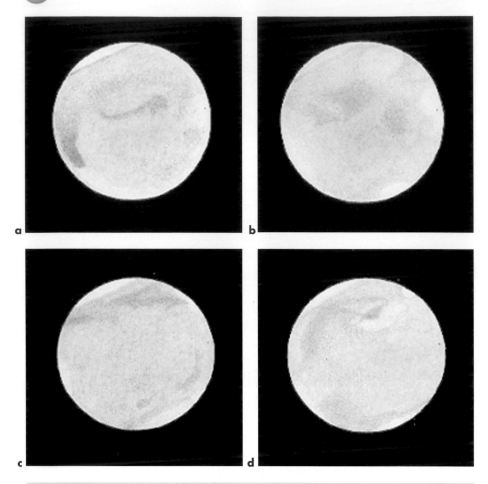

Figure 6.2. Drawings from one rotation of Mars: **a** Sinus Sabaeus/Sinus Meridiani region, 4 June 2001, 11 pm, 343×, seeing scale II–III; **b** Mares Tyrrhenum/Cimmerium and Syrtis Major regions; Hellas and Syrtis Blue Cloud evident, 16 June 2001, 12:50 am, 458×, seeing scale III; **c** Mare Sirenum/Elisium/Propontis regions, 20 June 2001, 10:15 pm, 343×, seeing scale II; **d** Solis Lacus/Mare Acidalium regions 4 July 201, 11 pm, 343×, seeing scale II.

Sighting of Mars' two moons are a stretch for the amateur at the best of times, even in ideal circumstances. They can be seen with a blocking bar to blot out the glare of the planet itself in the field of view, though from city confines, I believe the moons will be too big a challenge for most observers. Again, as with most things, there is no reason why you shouldn't try, but do not for one minute be tempted to try it with your image intensifier!

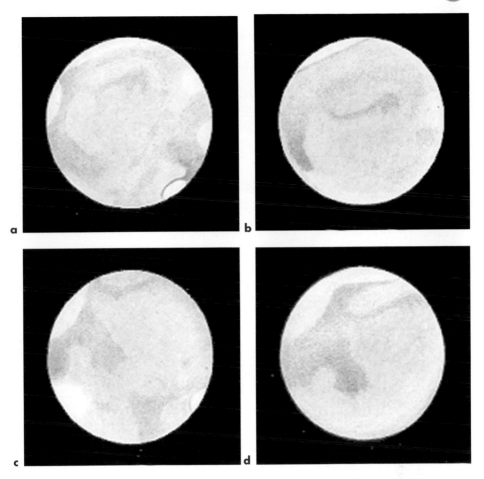

Figure 6.3. a, b Two examples of the Sinus Sabaeus/Sinus Meridiani regions:
a 19 May 1999, 9:30 pm, 458×, seeing scale II–III; **b** 4 June 2001, 11 pm, 343×, seeing
scale II–III. **c, d** A similar comparison from two separate oppositions (Syrtis Major region):
c 23 May 1999, 9:30 pm, 458×, seeing scale III; **d** disk is in partial phase, 5 May 2001,
1:45 am, 343×, seeing scale III.

Mapping Mars

At the end of the best period of an apparition, it is
extremely satisfying to produce a Martian map, incorporating the delicate shadings and colors of what you
have observed and recorded on your full-disk drawings. In order to facilitate this, and remembering that
we are not fooling ourselves that we are producing
something of unique value, make a blank projection

from a reputable existing map. This will indicate the correct Mercator's projection of latitude and longitude; for our needs, there is no need to reinvent the wheel. Examining the existing map and taking note where the features fall, arrange your full-disk drawings in appropriate sequence. Place a light outline of the equivalent observed features in the same places on the map. Draw the features as you saw them; they may well differ from photographs or other maps, so try to recreate exactly the look of your full-disk drawings on the map itself. We are making something akin to an orbital view of the entire surface (or at least as much as has been shown to us), and with the perspective of our own eyes and judgments.

You will be able to produce quite a serviceable flat projection in the most straightforward way with this method. It can combine the best detail from all of your individual observations over an entire opposition, and also serves as a great cross-reference for anyone examining your full-disk drawings. It is, however, quite time consuming. Of course, changing cloud detail and limb hazes should be excluded. All in all, not a bad result for mere city-bound observations!

The map reproduced in Figure 6.4 was based on my observations of the apparition of 1999, April through June.

Jupiter

This planet, the king of our planetary system, will reward you endlessly at each successive apparition, and is arguably the best single sight we have to observe from the confines of our suburban surroundings (Figure 6.5). It is surprising how much the way we see it has been altered forever by the close-up views we have had from visiting spacecraft. Looking back at the work of past generations of observers, it becomes clear how very differently their eyes perceived the planet. Not all of the differences can be attributed to Jupiter's ever-changing nature. At least some of these, along with descriptions of old, do not seem to tie in with so many of the visual attributes that seem so obvious to us now. Examine any literature and drawings from before the space age (such as Peek's *The Planet Jupiter*) and this will become clear, and it is surprising to me how little all of this is commented upon. However, as past

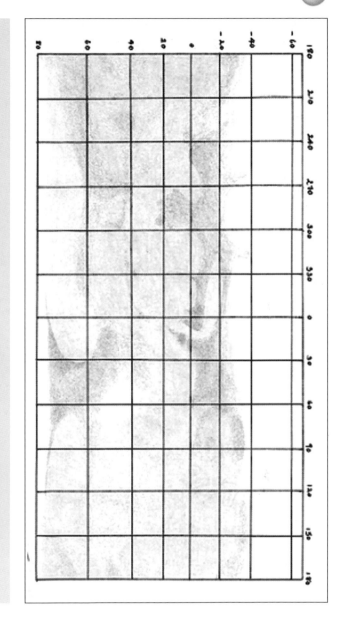

Figure 6.4. Map of Mars 1999.

observers practiced, for the best preliminary visual study it is wise to make full-disk drawings of the planet. As with Mars, I make up large numbers of blanks on a black background which I use for this purpose. It is not enough to produce a circular disk on a black background, since Jupiter has pronounced flattening at the poles. First of all, cut out an accurately shaped white "disk" and glue it to a black square. This

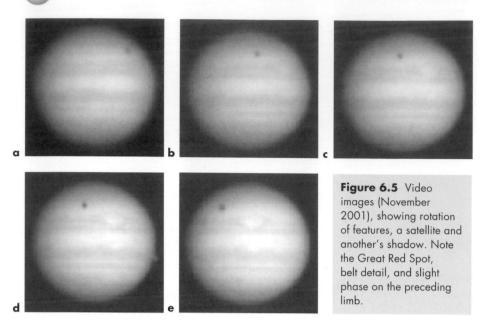

Figure 6.5 Video images (November 2001), showing rotation of features, a satellite and another's shadow. Note the Great Red Spot, belt detail, and slight phase on the preceding limb.

template can be used for any number of copies. You will not need to concern yourself with phases on your blanks as they are virtually non-existent on Jupiter, appearing only as subtle limb darkenings. Jupiter is so large and detailed that it responds best with medium–high magnifications. Little additional detail is likely to be seen with the highest powers, and the image of the planet will deteriorate noticeably from the beautifully defined images we see at somewhat lower powers. I get very good results alternating at 229× and 343×, and occasionally 458×, depending on conditions, as well as the point I have reached during the specific observation. The lower power produces the most beautiful looking images, whereby the higher ones are useful for resolving additional detail within a specific feature.

As with Mars, certain filters can help in making detail stand out. I find that blue filters can produce some of the best contrast results; the reddish hues of some of the belts, and especially the Great Red Spot, seem more striking. I have never been particularly taken with the use of most other filters on Jupiter, an object that stands up pretty well without any help at all. You will notice some darkening of the planet's limbs, due to its gaseous nature and reflected light being absorbed, although this is not nearly so obvious as some descriptions would have you believe. The most prominent features, with the greatest color and contrast, lie nearest to the equatorial regions. Here it is quite normal to see the belts in striking shades, while so-called festoons, wisps and loops in the equatorial zone are common. Within the two main belts may be many subtle colors and zones, which, depending on the aperture of your telescope, can be detected, along with something of the swirling nature of these regions as revealed by spacecraft. More temperate regions, and those within the polar regions are decidedly vaguer and paler. It is here that white spots can become prominent, and sometimes even dark streaks, approaching shades of black, are superimposed. For the most satisfactory drawings of the planet it is good to adopt the following procedure:

- Unless there are unusual features immediately apparent, it is wise to wait a while before starting to sketch. You should take the opportunity at this time to determine, as accurately as possible, the correct latitudes of the cloud-belts and lightly indicate them on the blank disk. Beware of the common error of placing the two equatorial belts too far apart, or making them too wide; I am all too aware of my own tendency to do these things. Also guard against making the features too big. Many of the same techniques already discussed in drawing Mars apply equally here, but specific approaches needed for Jupiter will rapidly become apparent.

- Soon your eyes will begin to adjust, moments of good seeing will likely have occurred, and you will really begin to see what there is to see. With a little more time the markings will become even clearer to you, and those now on the following side, that you first became aware of, will have moved closer to the meridian. Jupiter's rapid rotation is noticeable almost immediately. As the features work their way across the meridian, it will be possible to have fully grasped their appearance, and the next set of features will be moving into view. This is a good time to start the drawing. In this way, features centering around both sides of the meridian will have been recorded with the best possible chance of accuracy, and centrally located features, having moved to the preceding limb, will also be best appreciated in relation to others. (Remember that since an educated eye sees best, these receding features will be more clearly understood as they move to the less favorable position.)

- The belts present a variety of subtle color contrasts. It is best to start by drawing the belts in gray, gradually evolving their detail and colors. The equatorial belts will likely always be the most colorful, with hues ranging from browns to ruddy, even brick reds; within these belts there is usually a wealth of detail, and complex dark and light streaks and blotches. Within the equatorial zone, you will see most of the festoons, loops and wisps so often written about. These are generally light blues and grays. Where they join the belts, they are often quite a deep bluish color. You will notice that these features have an independent rotation, and changes of relative positions to the belts themselves occur from day to day. Jupiter's other belts are less prominent, as well as being narrower, and usually more likely to take shades of gray in coloration. Always be open to what you see; Jupiter can undergo frequent and surprising changes in any region.

- The Great Red Spot is quite a feature in itself, ranging from salmon pink, with a red center, typical in recent years, to any number of other variants. It has been fainter than normal over this time, and it is here that the aperture of my telescope pays big dividends. There has never been the slightest difficulty in making out the Spot or its color, which is likely to appear much grayer in smaller scopes. Watch for swirling and disturbed activity where it is superimposed on the south equatorial belt. Much of the time, these disturbances occupy most of the circumference of the belt, in varying degrees. As for the bright yellow/white zones between the belts, look carefully: there is more variation of color here than is at first obvious. CCD imaging reveals more of this than does the eye, which is easily overwhelmed by the brightness of the planet. I still know of no way to fully represent by drawing the color and "look" of these zones on the page, and have had to settle for something of an approximation. It is as if the predominant color has a luminosity of its own, like a light shining from behind, that cannot be represented by drawing, CCD image, or photograph.

- In order to have the best chance of seeing and recording all that is there, as an additional step it seems to me to be acceptable to first determine features, assign positions, and outline them on the following limb. Then wait for the features to move more centrally, for better clarity before completing them. You will have to make careful adjustments, as the perspective is altered as the feature moves from more of an edge-on position to a central one. This step does, of course, produce a drawing even more aligned with time exposures, and you may wish to draw this limb based only on what you know at the time.

However you proceed, such drawing endeavors will prove an invaluable learning experience in the ability to see all that is there. The reproductions here can never reproduce the subtleties of the original drawings, and CCD images may well at times record even more detail. However, I have seen very few planetary CCD images that have anything of the striking sharpness, subtleties and impression of the eye's own vision, although I have often been astounded with the wealth of detail that really fine CCD examples can reveal.

The examples of full-disk drawings in Figure 6.6 exemplify the essential appearance of the planet, as seen with my 18-inch reflector.

More satisfying than merely recording full-disk drawings of Jupiter is focusing on individual features or regions. This can be a small detail or even a large and turbulent feature such as the Great Red Spot and its surroundings, which can change perceptibly over a very short time. Over multiple occasions these ever-changing aspects can be much better represented in this way than by other means I know. The greatest bonus of all is that you will see much more detail than you otherwise might, with only a limited region to focus your attention on. I have included examples of this kind of regional strip drawing in Figure 6.7; it is possible to elect a scale that would be altogether too large for a full-disk drawing.

Additionally, the Galilean satellites provide some opportunities for amateur observation. The motions between planet and satellites are of great interest to many observers, with transits, shadows cast on the planet itself, and disappearances and re-emergences from behind the planet or its shadow always an appeal-ing topic. With sufficient aperture, best seeing condi-tions, and the highest magnifications practical (even up to 680× in my case), it is possible to glimpse tantalizing snippets of detail on the moons, most notably on Ganymede, the largest of them. While at least 25-inch aperture or so is necessary for pursuing effective surface studies of Jupiter's moons, I have been able to note tantalizing shreds of detail with my 18-inch at

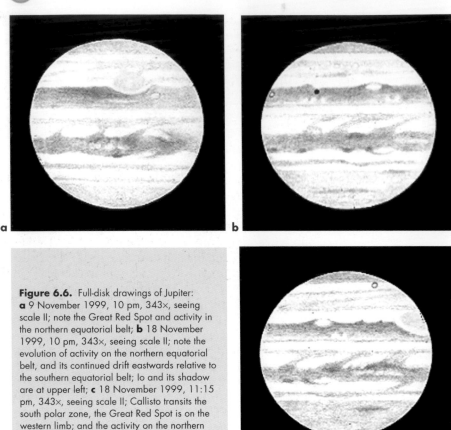

Figure 6.6. Full-disk drawings of Jupiter: **a** 9 November 1999, 10 pm, 343×, seeing scale II; note the Great Red Spot and activity in the northern equatorial belt; **b** 18 November 1999, 10 pm, 343×, seeing scale II; note the evolution of activity on the northern equatorial belt, and its continued drift eastwards relative to the southern equatorial belt; Io and its shadow are at upper left; **c** 18 November 1999, 11:15 pm, 343×, seeing scale II; Callisto transits the south polar zone, the Great Red Spot is on the western limb; and the activity on the northern equatorial belt has evolved as the belt has drifted east relative to the Great Red Spot.

times of superior viewing, and I understand that lesser apertures have also shown striking details at times. You will need to wait for transits, with the satellite viewed against the planet itself to be successful; otherwise the contrast against the dark sky will prove overpowering.

Cylindrical Projections

We obviously cannot make maps of Jupiter, with its constantly changing face. Instead, and actually more interesting to me than other forms of Jovian drawing, are cylindrical projections of the planet in rotation. You may have already seen photographic versions of these relayed to Earth from spacecraft. It is even possible to capture a complete rotation when conditions

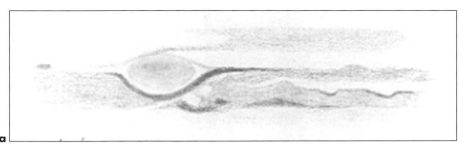

a

b

c

Figure 6.7. Drawings of regions of Jupiter: **a** 21 December 2000, 8:30 pm, 343×, seeing scale I: The Great Red Spot. Contrasted examples of activity in one region (virtually the same portion of the northern equatorial belt) over a short combined time frame: **b** 21 December 2000, 7:30–10 pm, 343×, seeing scale I; **c** 23 December 2000, 7:50–10:45 pm, 343×, seeing scale II–III.

allow, and each feature can be scrutinized as it crosses the central meridian, a real advantage over other methods of observation. It is true that this will tie up your telescope for a long session, but it is so rewarding that it now has become my preferred method for observing and drawing Jupiter. In any given apparition, it is really only instructive to make a few full-disk drawings for the general appearance of the planet, spending more time on recording limited regions or features, and especially these cylindrical projections. It is an infinitely more informative and satisfying process. Nevertheless, it is not wise to commence a cylindrical projection until you have had some experience with full-disk drawings first. Because you will have familiarized yourself with the procedures necessary to recreate the appearance of the planet, the process and end result will likely be far less frustrating.

A cylindrical projection takes the disk and stretches out the polar regions so that the sphere of the planet can be laid out flat. This will involve expanding all features increasingly in regard to their positions north and south of the equator, sort of a map with Mercator's projection in latitude only. It is not nearly as difficult a procedure as it sounds, as we will be drawing only what we see on the central meridian at any given time. By staying with that principle, these features will stretch themselves out automatically for you. Remember to let them expand laterally, ignoring their actual longitudinal appearance at the moment; just draw what you see close to the meridian itself. Usually the most striking details occur on the two equatorial belts, and because they do not involve significant north and south latitudes, they will be quite easy to record in correct lateral proportion.

The procedure I use is as follows: To begin, use a long sheet of paper, such as 17×11 inches. Determine ahead of time the scale you are going to use, incorporating hourly rotation. Draw markers vertically on this sheet to correspond to this hourly rotation. This will keep the proportions correct as long as you always note the time you are observing. Remember to establish the correct spacing of the belts relative to the scale of the drawing. Once you are under way, you can also utilize these vertical markers, and the spaces in between, to note the time that specific features cross the central meridian, if you wish to make them part of the record. Having completed the "draft" copy, just as with full-disk drawings, begin a color final version. Draw and fill in the features, then add detail in exactly the same manner as before. Figure 6.8 shows a cylindrical projection from the 2000/2001 apparition.

Saturn

What a sight this is in the telescope! In many ways there is nothing to rival the real time appearance of this planet, as it seems to hang in space like a luminous flying saucer (Figure 6.9). It responds very favorably to CCD video imaging, but you will probably find that soon you want to try to capture the additional subtleties that you see in the eyepiece. By comparison to video, drawing it is another matter entirely, as Saturn is amongst the greatest challenges to our drawing skills that we are likely to face. Additionally, satisfactory blank disks of Saturn are extremely hard to produce, the greatest difficulty being the ever-changing tilt and aspect of the rings. The Crepe Ring and its shadows on the disk only compound this further. You might use a current video image, blanked out and enlarged to appropriate size as a template. I have found a satisfactory solution has been to draw the planet and rings freehand, with accurate proportions being established at the outset. This can easily be accomplished by

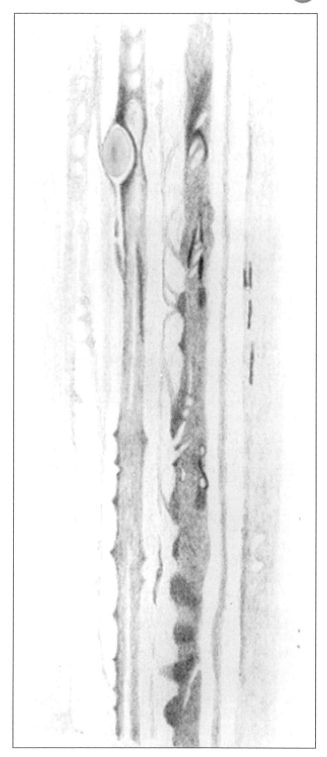

Figure 6.8. Jupiter 2001: 6 January 2001, 6–11:30 pm, 229× (blue filter)/343× (no filter), seeing scale II.

Figure 6.9. Video image of Saturn, November 2001. The rings are almost fully open, with shaded zones on the rings. Note the prominent belt and two others, the dark polar region, the absorption zone made by the Crepe Ring, and almost perfect alignment of the planet relative to the Sun and Earth, since there is only a slight wedge of shadow from the disk on the rings themselves.

marking on the paper small dots at the outermost and innermost places (check these dimensions from any photograph or video image), and then outline the body and rings of the planet itself to correspond to the planet's appearance at the time. Since I don't draw Saturn more than once or twice a year (for obvious reasons to anyone who tries it), I can afford to let each drawing be more time-intensive than would normally be the case with other subjects.

Instead of an entirely black background extending to the corners of the page, I use a black pencil on white paper to suggest black space around the planet. This consists of dark shading outlining the planet and rings, continuing outwards but gradually reducing its intensity until it blends into the white of the paper. It is also drawn with linear motions parallel to the plane of the planet, for a clearer effect that does not compete with the orientation of the rings. Having done this, capturing the coloring and relative brightness of the rings should not pose too much of a problem, and the disk itself should be a "breeze" after the challenges of Jupiter.

The sheer beauty of this planet is mesmerizing, in ways only live viewing can provide. The overall color is something of a deep yellow-gold, and even small telescopes will put on a worthy show. The variations of hues on Saturn are less obvious than on Jupiter, and the changes from month to month are far slower, and may not be detectable for much longer periods. The number of cloud belts visible also will be fewer than on Jupiter, and the contrast between belts and zones more subtle. These cloud belts do not normally exhibit such turbulent variations and detail as do Jupiter's, but you should be able to detect a typical equatorial belt redness versus distinct grays of the polar regions, and sometimes entire regions may be lighter than the surrounding area. Spots are more unusual, and certainly harder to see, although there have been some notable sightings over the years. Perhaps the most famous of these was the Great White Spot discovered by British comedian Will Hayes in the 1930s, and long since faded into history.

Look for variations in the entire ring system itself; you will see areas of light and shade. The Cassini Division is an easy mark; not so the infamous Encke Division, which is often confused with a dark zone within the outer A-ring, closer to the planet than the Encke feature. It took spacecraft to confirm its existence undeniably since it was too fine to show itself on photographs before the time of space exploration. As far back as the nineteenth century, certain noted observers also reported seeing structures on the rings themselves, which seems to suggest that they did in fact see traces of the famous "spokes" shown in views obtained by much more recent spacecraft. At the time these could not be confirmed by other astronomers, but now it would appear there was something to these old observations. Today, with firm confirmation of their existence, some amateur observers also have claimed to have discerned suggestions of the "spokes", eye preparation again revealing things previously unseen and unknown. As of yet, I cannot join their ranks, but maybe you will. With little more than small apertures the Crepe Ring becomes possible to see; its transparency and bluish-gray color become more apparent with increasing aperture (probably anything 8 inches or more) and darker immediate surroundings. Be sure to baffle your backyard suburban sight against light intrusions. Most good telescopes above 6 inches will reveal the Crepe Ring, though in larger apertures it is very striking, its subtle coloration and appearance likely to require these larger instruments.

Unless the seeing is at its best, any fine details on Saturn are notoriously difficult to detect, although the total image of Saturn, including its rings, appears quite sizeable and about the equal of Jupiter. The actual disk is very much smaller, approximately a third of the whole. You will find, however, that this planet is quite capable of withstanding some very high magnifications on good nights, far higher than with most other objects. On one notable evening many years ago with my old $12\frac{1}{2}$-inch reflector I was able to use the astounding magnification of 720×, with the greatest clarity and ease!

The main drawing in Figure 6.10 was made on a particularly good night when the air was almost still; the planet was placed at one of its most beautiful positions relative to us, with an almost fully open ring system for us to see. It is a good example of an entirely freehand drawing, made according to the guidelines previously discussed. The complexities also explain why I don't draw Saturn too often! Don't be misled by my drawing; although it is a very close representation of the visual appearance of Saturn that night, the so-called Encke's Division near the outer edge of the ring system, at 0.1 minute of arc, is actually beyond the diffraction limit of even an 18-inch telescope such as mine. However, I do believe on the night that I drew the planet, I did indeed glimpse something of it, as has been the case with other observers viewing similar high contrast planetary features. I don't believe I mistook it for the dark feature within the A-Ring. Encke's Division is so routinely and glibly described by so many of today's amateurs that you will begin to think that you or your telescope

Figure 6.10.
Drawing of Saturn: 28 November 1999, 10:45 pm, 343×, seeing scale II.

are at fault if you have failed to see it. It is not likely that most of these claims can be substantiated, as very few amateurs possess instruments of sufficient aperture to reveal such a fine feature. It is difficult even with 40 inches, and certainly requires excellent seeing. Here again we can see how the eye has a better chance of seeing something, once it knows how it should appear. But remember, there is a fine line between seeing and convincing ourselves we see something which we can't. As always, be vigilant for the truth.

The aspect of the rings relative to us is affected in exactly the same way as are the variations in Mars' appearance from year to year. The changes we perceive are affected by the axial tilt of the planet relative to us. It has nothing to do with the rings themselves "opening and closing"! This remarkable phenomenon of our relative planetary positions occurs over approximately a 15 year cycle, and produces incredible variations in the appearance of the planet, not only in the angle that the rings present themselves, but also in shadows projected on them by the planet itself.

In order to show the appearance of the rings in a largely edge-on position relative to our line of sight, it was necessary to utilize a drawing made long ago with my old $8\frac{1}{4}$-inch reflector (Figure 6.11). I simply have not had the 18-inch enough years to provide the range that the rings show us. However, accessing drawings made long ago only illustrates one of the great pleasures to be enjoyed by having a collection of images made over many years. (This old drawing, from 1967 does not, of course, exhibit the resolution that I enjoy these days, and was also made in black-and-white.)

Figure 6.11.
Drawing of Saturn: 22 October 1967, 10:15 pm GMT; 200× $8\frac{1}{4}$-inch reflector, seeing scale II. The rings are almost edge-on, the Crepe Ring is just detectable, and the planet has a bright equatorial zone.

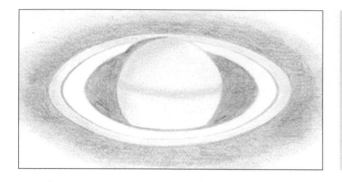

Figure 6.12.
Drawing of Saturn: 16 April 1974, 9:10 pm Eastern Time; 167× 3-inch refractor, seeing scale I. The rings are fully open, and there is a shadow on the pre-ceding side; compare with Figure 6.9.

Also possibly of interest to you is another drawing (Figure 6.12) made through a smaller telescope yet (a 3-inch refractor I had for some years). Both it and the drawing above show just how well these smaller apertures can perform on planetary subjects. Although detail is noticeably less with both of these drawings than through the large scopes available to amateurs these days, the breathtaking nature of this almost surreal subject still shows superbly. It will indicate just how successful you will be with practically any aperture you might turn on it.

A comment on viewing the satellites: unlike those of Jupiter, which, with large enough apertures occasionally provide us opportunities to glimpse surface detail, those of Saturn will not grant any such favors to us. The majority are simply too far away and small to see as scarcely more than star-like points. Many observers nevertheless enjoy witnessing the interplay between the visible satellites themselves, the planet and its rings.

Saturn is probably the crowning glory of our short tour around the best of the Solar System, but it is now time to move on to the great expanse beyond. This, of course, is the realm so often written off for suburbanites as being hopelessly ruined by our cities. While I aim to show how we may access some of that lost realm and what part of it is available to us, we should always take advantage of the Moon and planets whenever they are observable in the sky. All we need is steady air, a clear enough horizon, and the curiosity to explore.

Chapter 7
Deep Space – Visual Impressions and Expectations: The Primary Catalog

The listing of deep space objects featured and described later in this chapter represents, in my view, the finest sights available to the Northern Hemisphere suburban dweller. Many are visible from the Southern Hemisphere as well. Almost without exception, few of these sights will be in the realm of the spectacular in the absence of any image enhancing equipment as described earlier. Because of the nature of them, we will do best on nights of best transparency, just as we would at a dark sky site. I have included real time illustrations of all of the objects themselves individually, as well as some comments on observing them. Generally, I leave detailed scientific and historical background about these marvelous sights to the many excellent reference sources that already exist; it has never been my intention to repeat material so readily available. I think it is also important to point out that from the suburbs it is usually not realistic to expect to see the kinds of extensive subtle detail one would observe from prime dark sky sites, so painstakingly described in many other volumes on the deep sky. This is not to say that the objects will not appear stunning or detailed; but in all fairness to you and your expectations, I will not pretend to see in them more than I have been able to. It is instead my hope that I can provide you with a realistic expectation of what you may be able to see from your city-bound backyard, and how these objects might appear to you.

Next to each listing, I have included some relevant information, depending on the type of object. These statistics (with the exception of the coordinates, updated to present values) are according to *Burnham's Celestial Handbook*. This source seems to me to be the most reliable source in its assessments. Others often seem overly optimistic regarding estimated visual magnitudes. Anticipating unrealistic object magnitudes may be quite disappointing for observations based on such estimates; Burnham seems much closer to the mark. However, just what you will see depends on a number of factors: your telescope's aperture, magnification used, clarity of sky, whether an image intensifier or light filter is employed, etc., the quality of all those things – plus, of course, your location. In considering these various factors, and the differences with your actual observations of the objects listed below, the answer is not as obvious at first as it may seem. The lowest magnification of a larger aperture will be much greater than that of a lesser aperture, but in illumination of the field, at any rate, the advantages of the larger size will be somewhat negated by the light diminution of its higher minimum power. So although the resolution and image size will not be of the same order with a smaller aperture, the results with most of the objects in this chapter should be pleasing with moderate-to-large telescopes in less than favorable conditions. Your visual expectations should not be too far out of line with what is shown in the illustrations, and they should serve as a reasonable guide.

All video images in this chapter were taken with the set-up described in Chapters 2 and 4, with my 18-inch reflector. Similarly, the drawings were made directly from observations (either with intensification, a light filter, or no added means at all), as outlined in the main listing. Be sure to experiment with everything available to you and not just according to my recommendations, since these reflect my own circumstances and preferences only. However, these recommendations will at the very least serve as a reliable basis for your own viewing methods. The deep space illustrations are also of several types: direct real time images from the image intensifier via a CCD video camera, drawings from observations with the image intensifier, drawings as observed through a narrowband light filter (and occasionally without), and a few others representing the combined separate views through filter and then intensifier. I do not specify the power used, but usually the image scale approximates the effect of what I experienced at the time using low-to-moderate powers. In my own case, this usually corresponds to around 80× or 160×. High power was rarely more than 240× for deep space subjects. (Occasionally, when a higher power was used, it is noted as such.)

The reality of "seeing" is seldom as obvious as you might anticipate. The images, particularly with some galaxies, will sometimes not be as bright as you might expect. In time, this ceases to be a factor, as your mind and eye will make their own relative adjustments. On the whole, you will notice a large gap between their grand observatory portraits and what is presented in the real time views; conversely, you may also be surprised when the reality comes close to the best photographs. When what you are seeing seems less obvious, don't be led astray. With perseverance, knowledge of what is really there, as well as what to expect, you will marvel at being able to discern so much, including so many subtle traces of all you know to be present. These subtleties will look just as good to you as if you saw their fully resolved being. You will share in the same wonder of real time viewing that has inspired me all these years; it is as if you have some direct contact with other places across the vast distances of space. Amazingly, some of this even holds true when viewing live on the TV monitor, which should be good news for those using Astrovid's StellaCam EX or SBIG's STV. And to think we are experiencing all of this in surroundings that have been deemed unfit for such observations by most serious observers!

There really is no shortage of grand deep space sights for us to enjoy from the city. In the individual descriptions that follow, I have included some visual information to guide you in this "most successful sights" primary list. The real time deep space images are presented without making them appear superior to the way they appeared to me at the time, as viewed directly through the eyepiece, intensified or otherwise. Most were made (drawings included) in the typical, mediocre suburban sky conditions that I experience the majority of the time from my location, and with which this writing is primarily concerned. Any adjustments to the video images were made only when it was necessary to make the contrast closer to that of the view through the eyepiece. Aside from this, the only allowance you might grant is in the slightly "digital" and less than perfectly defined appearance of some of these images (particularly in the enlarged examples). One has to remember that they are simple video images after all, basically unprocessed, and remarkable for having been captured in real time. In a straight comparison, image to image, obviously they will not usually compare with time exposure photography or CCD imaging. The crispness, brilliance, and detail of the live view, either direct through the image intensifier or to some degree on the video monitor, is also impossible to transfer to the page; the printing process reduces this further. However, these video snapshots, in conjunction with drawings provided wherever possible, will provide you with a remarkable degree of preparation for your own real time viewing.

The Most Successfully Seen Objects: Our Primary List

The enhancing device listed with each object and abbreviated as below, likely indicates the best way to view it:

INT = Image Intensifier (best in conjunction with a Barlow lens)
NBF = Narrowband Light Filter (lowest power)
§ = Unaided Viewing (no intensifier or filter)

NGC 40   magnitude 10.5, angular size 60″ × 40″, celestial coordinates (00130n7232), constellation **Cepheus**. Planetary nebula; prominent ring with subtle ansae and 11.5 mag. central star. INT.

NGC 224 (M31)   magnitude 5, angular size 160′ × 40′, celestial coordinates (00427n4116), constellation **Andromeda**. Sb galaxy; the nearest galaxy, extremely large and bright but not resolvable visually. INT. In field with:

NGC 205   magnitude 10.8, angular size 8′ × 3′, celestial coordinates (00404n4141), constellation **Andromeda**. E6 galaxy. INT. Also in field:

NGC 221 (M32)
magnitude 9.5, angular size 3.6′ × 3.1′, celestial coordinates (00427n4052), constellation **Andromeda**. E2 galaxy. INT.

NGC 253
magnitude 7.0, angular size 22′ × 6′, celestial coordinates (00476s2517), constellation **Sculptor**. Spectacular large Sc galaxy. INT.

NGC 598 (M33)
magnitude 6.5, angular size 60′ × 40′, celestial coordinates (01339n3039), constellation **Triangulum**. Sc galaxy; hint of spiral structure, bright nebula NGC 604 visible. NBF.

NGC 650 (M76)
magnitude 11, angular size 140″ × 70″, celestial coordinates (01424n5134), constellation **Perseus**. **Little Dumbbell Nebula**, planetary nebula of irregular shape. NBF.

NGC 869/ 884
magnitude 7 each, angular size 35′ each, celestial coordinates (02190n5709/ 02224n5707), constellation **Perseus**. **Sword Handle Double Cluster**; two open clusters. INT/NBF/§.

NGC 1068 (M77)
magnitude 10, angular size 2.5′ × 1.7′, celestial coordinates (02427s0001), constellation **Cetus**. Sb galaxy: bright, hints of structure. INT.

NGC 1097
magnitude 10.6, angular size 9′ × 5.5′, celestial coordinates (02463s3016), constellation **Fornax**. SBb galaxy; some detail and structure visible. INT.

IC 418
magnitude 8, angular size 14″ × 11″, celestial coordinates (05275s1242), constellation **Lepus**. **Spirograph Nebula**; very bright oval planetary nebula with 11 mag. star. INT.

NGC 1952 (M1)
magnitude 9, angular size 5′ × 3′, celestial coordinates (05345n2201), constellation **Taurus**. **Crab Nebula** supernova remnant; resolution of tendrils possible with sufficient aperture. NBF.

NGC 1976 (M42)
magnitude 5, angular size 65′, celestial coordinates (05354s0527), constellation **Orion**. **Great Nebula in Orion**; emission nebula. NBF/INT/§(exceptional); INT emphasizes Huygenian region with stars.

NGC 2261
magnitude 10, angular size 2′, celestial coordinates (06392n0844), constellation

Monoceros. Hubble's Variable "Fan" Nebula; possible emission or reflection nebula with fan-like shape and variable outline. INT.

NGC 2359 magnitude 11, angular size 6′ × 8′, celestial coordinates (07186s1312), constellation **Canis Major. Thor's Helmet**; emission nebula. NBF (reveals nebula)/INT (reveals stars).

NGC 2392 magnitude 8, angular size 40″, celestial coordinates (07292n2055), constellation **Gemini. Eskimo Nebula**; planetary nebula with 10 mag. central star. High-power §, or INT.

NGC 2440 magnitude 11.5, angular size 50″× 20″, celestial coordinates (07419s1813), constellation **Puppis**. Planetary nebula of complex appearance, with 16 mag. central star and lobes, in some ways like a small Saturn. INT.

NGC 2683 magnitude 10.6, angular size 9′ × 1.3′, celestial coordinates (08527n3325), constellation **Lynx**. Sb galaxy, almost edge-on. INT.

NGC 3034 (M82) magnitude 9.2, angular size 8′ × 3′, celestial coordinates (09558n6941), constellation **Ursa Major**. Irr. galaxy; exceptional, dark lanes, mottling. INT.

NGC 3115 magnitude 10, angular size 4′ × 1′, celestial coordinates (10052s0743), constellation **Sextans. Spindle Galaxy**; bright E7/SO galaxy. INT.

NGC 3132 magnitude 8.2, angular size 84″ × 52″, celestial coordinates (10069s4021), constellation **Vela. Eight-Burst Nebula**; planetary nebula similar to the Ring Nebula, with 10 mag. central star. INT.

NGC 3190 magnitude 12, angular size 3′ × 1′, celestial coordinates (10181n2150), constellation **Leo**. Sb galaxy; edge-on, dust lane. INT. In field with:

NGC 3193 magnitude 12, angular size 0.9′ × 0.9′, celestial coordinates (10184n2154), constellation **Leo**. EO galaxy. INT. Also in field:

NGC 3187 magnitude 13, angular size 1′ × 0.3′, celestial coordinates (10178n2152), constellation **Leo.** SBc galaxy; at times just visible. INT.

NGC 3242   magnitude 8.9, angular size 40″, celestial coordinates (10248s1838), constellation **Hydra. Eye Nebula;** planetary of startling appearance. INT.

NGC 4111   magnitude 11.6, angular size 3.4′ × 0.8′, celestial coordinates (12071n4304), constellation **Ursa Major.** E7 galaxy; edge-on, striking. INT.

NGC 4406   magnitude 10.5, angular size 3′ × 2′,
(M86)   celestial coordinates (12262n1257), constellation **Virgo.** E3 galaxy at the **Center of the Virgo Cluster;** strong red spectrum; small elliptical galaxy nearby. INT. Near:

NGC 4374   magnitude 10.5, angular size 2′ × 1.8′,
(M84)   celestial coordinates (12251n1253), constellation **Virgo.** E1 galaxy, plus two additional edge-on galaxies. INT.

NGC 4565   magnitude 10.5, angular size 10′ × 3′, celestial coordinates (12363n2559), constellation **Coma Berenices.** Sb galaxy; famous edge-on galaxy with exceptional, prominent dust lane. INT.

NGC 4594   magnitude 8.2, angular size 7′ × 1.5′,
(M104)   celestial coordinates (12400s1137), constellation **Virgo/Corvus. Sombrero Galaxy;** Sa/Sb galaxy with exceptional, dark equatorial lane. INT.

NGC 4736   magnitude 8.9, angular size 5′ × 3.5′,
(M94)   celestial coordinates (12509n4107), constellation **Canes Venatici.** Sb galaxy; very bright and compact. INT.

NGC 4826   magnitude 8.6, angular size 7.5′ × 3.5′,
(M64)   celestial coordinates (12567n2141), constellation **Coma Berenices. Black Eye Galaxy;** Sa galaxy, "black eye" visible. INT.

NGC 5128   magnitude 7.2, angular size 10′ × 8′, celestial coordinates (13255s4301), constellation **Centaurus.** SO/pec galaxy; round, detail and central band visible, exceptional. INT.

NGC 5139   magnitude 4, angular size 30′, celestial coordinates (13268s4729), constellation **Centaurus. Omega Centauri;** astounding globular cluster, finest known. INT/§.

NGC 5236   magnitude 8, angular size 10′ × 8′, celestial
(M83)   coordinates (13370s2952), constellation
           Hydra. Sc galaxy; strong emission spectrum in nucleus. INT.

NGC 5272   magnitude 6, angular size 18′, celestial
(M3)   coordinates (13422n2823), constellation
           Canes Venatici. Globular cluster; beautiful, well-resolved to center. INT.

NGC 5746   magnitude 11.7, angular size 6.5′ × 0.8′, celestial coordinates (14449n0157), constellation **Virgo**. Sb galaxy; edge-on, dust belt, bright condensations. INT.

NGC 5866   magnitude 11.1, angular size 2.9′ × 1′, celestial coordinates (15065n5546), constellation **Draco**. SO galaxy; elongated with prominent thin dust lane; exceptional. INT.

NGC 5904   magnitude 6.2, angular size 13′, celestial
(M5)   coordinates (15186n0205), constellation
           Serpens. Superb globular cluster. INT.

NGC 6205   magnitude 5.2, angular size 23′, celestial
(M13)   coordinates (16417n3628), constellation
           Hercules. Globular cluster; exceptional; look for "propeller" lanes. INT.

NGC 6266   magnitude 6.5, angular size 6′, celestial
(M62)   coordinates (17012s3007), constellation
           Scorpius. Globular cluster. INT.

NGC 6302   magnitude 9.6, angular size 2′ × 1′, celestial coordinates (17137s3706), constellation **Scorpius. The Butterfly** or **Bug Nebula**; possible planetary nebula of irregular shape, like a flattened figure "8". INT.

NGC 6341   magnitude 6.5, angular size 8′, celestial
(M92)   coordinates (171171n4308), constellation
           Hercules. Globular cluster; uneven distribution, smaller than nearby M13, but impressive. INT.

NGC 6369   magnitude 11, angular size 28″, celestial coordinates (17293s2346), constellation **Ophiuchus. Little Gem**; planetary nebula with perfectly circular ring and 16 mag. central star, easily seen with image intensifier in my 18-inch from my location. INT.

NGC 6402   magnitude 9, angular size 6′, celestial
(M14)   coordinates (17376s0315), constellation
           Ophiuchus. Globular cluster.

NGC 6514 (M20)	angular size 25', celestial coordinates (18023s2302), constellation **Sagittarius**. **Trifid Nebula**; emission/reflection nebula; exceptional; three dark lanes. NBF.
NGC 6523 (M8)	magnitude 5, angular size 80' × 40', celestial coordinates (1803s2423), constellation **Sagittarius**. **Lagoon Nebula**; emission nebula; exceptional; with cluster NGC 6530. NBF (INT reveals **Hourglass**).
NGC 6543	magnitude 8.6, angular size 22" × 16", celestial coordinates (17586n6638), constellation **Draco**. **Cat's-Eye Nebula**; planetary nebula; exceptional; helical structure partly resolved. INT.
NGC 6611 (M16)	magnitude 6.5, angular size 25', celestial coordinates (18188s1347), constellation **Serpens**. Open cluster, with **Eagle Nebula**. NBF.
NGC 6618 (M17)	magnitude 6, angular size 45' × 35', celestial coordinates (18208s1611), constellation **Sagittarius**. **Omega Nebula**; emission nebula; exceptional detail. NBF/INT.
NGC 6656 (M22)	magnitude 6, angular size 18', celestial coordinates (18364s2354), constellation **Sagittarius**. Globular cluster; exceptional; large, bright and resolved. INT.
NGC 6705 (M11)	magnitude 6, angular size 12', celestial coordinates (18511s0616), constellation **Scutum**. **Wild Duck Cluster**; impressive open cluster with dense, dark nebula near north. INT/NBF.
NGC 6720 (M57)	magnitude 9, angular size 80" × 60", celestial coordinates (18536n3302), constellation **Lyra**. **Ring Nebula**; planetary nebula; exceptional; subtle detail in ring visible, as well as central star. INT.
NGC 6826	magnitude 8.8, angular size 25", celestial coordinates (19448n5031), constellation **Cygnus**. **Blinking Nebula**; round planetary nebula with 11 mag. central star. § (very good at high power) or INT.
NGC 6853 (M27)	magnitude 8, angular size 8' × 5', celestial coordinates (19596n2243), constellation **Vulpecula**. **Dumbbell Nebula**; planetary nebula; exceptional. NBF.

NGC 6960/ angular sizes $70' \times 6'/60' \times 8'$, celestial
    6992   coordinates (20457n3043/20564n3143), constellation **Cygnus**. **Veil Nebula**; large emission nebula with filamentary structures. These are minor parts of the same nebula (of which NGC 6960 and NGC 6992 are the major components). Other components are NGC 6974/6979/6995. NBF.

NGC 7008   magnitude 12, angular size $85'' \times 70''$, celestial coordinates (21006n5433), constellation **Cygnus**. Planetary nebula; heart-shaped. NBF.

NGC 7009   magnitude 8, angular size $25''$, celestial coordinates (21042s1122), constellation **Aquarius**. **Saturn Nebula**; planetary nebula; exceptional. INT.

NGC 7027   magnitude 9, angular size $18'' \times 11''$, celestial coordinates (21071n4214), constellation **Cygnus**. Planetary nebula; star and two separate lobes on one side. INT.

NGC 7078   magnitude 6.5, angular size $10'$, celestial
(M15)   coordinates (21300n1210), constellation **Pegasus**. Globular cluster; exceptional, resolved, irregular. INT.

NGC 7089   magnitude 6, angular size $7'$, celestial
(M2)   coordinates (21335s0049), constellation **Aquarius**. Globular cluster; outstanding, resolved; look for dark lane near northwest corner. INT.

NGC 7662   magnitude 8.5, angular size $32'' \times 28''$, celestial coordinates (23259n4233), constellation **Andromeda**. Planetary nebula; exceptional, detail. INT.

A Detailed Guide to Viewing These Objects

NGC 40

This planetary (Figure 7.1), a seemingly unlikely candidate for impressive viewing in suburban locations, shows up only moderately with light filters. It was one of the greater surprises when I turned my image intensifier upon it, and suddenly the encircling ring

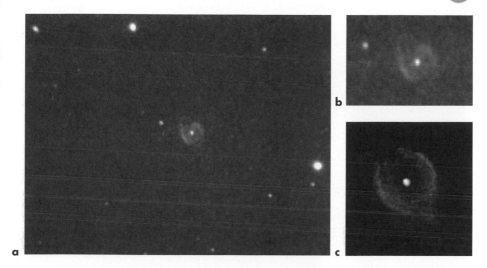

Figure 7.1. NGC 40: **a** video frame (high power): intensifier; **b** drawing: intensifier; **c** video frame: intensifier.

became very bright and prominent, together with subtle ansae features at each side; the central star blazed forth strikingly. It is one of the most impressive of all planetaries when viewed in this way, and much like its observatory portraits.

NGC 224 (M31) The Great Andromeda Galaxy

With its vast dimensions alone (being so close to us in space at a "mere" 2.3 million light years!), this object is one of the most imposing available to us (Figure 7.2). Surprisingly, in the telescope, it may at first prove to be a major disappointment; how could something so huge and bright still reveal so little of its true nature to us? While it is unfortunate that much of our first reaction is well founded, once we realize that it is not going to reveal itself fully to us in real time (particularly from our city locations), there are nevertheless numerous features to explore. With experience, you may be able to discern something of the large dust lane between its core and spiral arms, and possibly the bright southern star cloud so evident on photographs. If you arm yourself with a detailed chart, you also may be able to detect some of the globular clusters that encircle the galaxy,

a

b

Figure 7.2. NGC 224 (M31): **a** the nucleus of the galaxy; video frame: intensifier. **b** This image intensifier-color view of the core region of M31 is reproduced courtesy of W.J. Collins of Collins Electro Optics. It was taken from suburban Denver in a 6-second exposure using a digital camera, in conjunction with Collins' 7-inch Astro-Physics refractor, equipped with his company's I3 Piece image intensifier. The dust lanes mentioned in the text are clearly visible, providing some expectation of the live intensified view with low powers (especially using the Collins unit), from a location with low humidity and high sky transparency.

in the same way as do those in the Milky Way; these require patience and viewing skills, as they appear more like star-like points. The sheer size of this galaxy in the eyepiece view needs to be fully appreciated (far exceeding the field of view for most telescopes, even at low power), making the identity and whereabouts of its main companion elliptical galaxies NGC 205 and NGC 221 (M32), hard to fathom. NGC 205 is much further away from the central core of M31 than you may realize; because it is not as bright as NGC 221 it is also harder to detect. Far in the future, all three will collide with the Milky Way, in a great cosmic interplay that will forever change their shapes and even result in the obliteration of some.

NGC 253

One of the grandest galaxies in the sky, NGC 253 (Figure 7.3) was missed by Messier due to its low southern placement; however, many Northern Hemisphere observers will still be able to view it when sky transparency permits; haze at the horizon of my own location makes it challenging. Because of its great apparent size, it may not be possible to see all of it in larger telescopes' fields of view, not to mention the difficulties due to limited brightness away from the core. These factors certainly apply to my telescope and location, and the images presented here are therefore of the central region only. The bright spot near the core is actually a globular cluster! In the live intensified

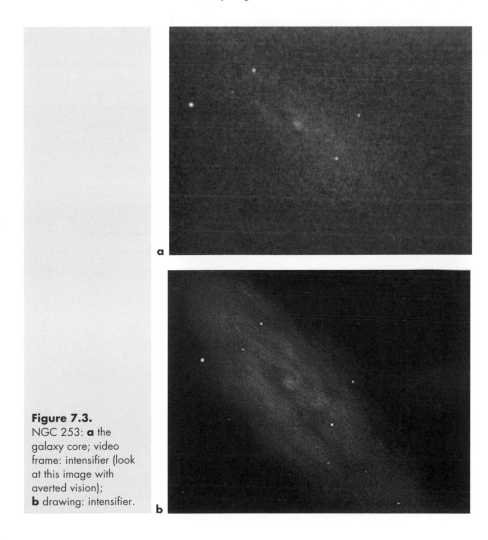

Figure 7.3.
NGC 253: **a** the galaxy core; video frame: intensifier (look at this image with averted vision); **b** drawing: intensifier.

a

b

view, traces of dark lanes and the region's mottled nature are detectable, remarkable to be caught at all by video in such conditions.

NGC 598 (M33) The Pinwheel Galaxy

This classic galaxy, probably the next closest to us after the Great Galaxy in Andromeda, seems to promise so much, but it can be amongst the most disappointing sights for which we may have high expectations (Figure 7.4). I know of observers who simply fail to see the "pinwheel" under any conditions. Knowing therefore that it tends to be elusive, you should only try to seek it out when the skies are at their most transparent. Nevertheless, it is possible to see it with some clarity in the problematic sky conditions we face most of the time. To spot it initially, you may do best with binoculars, though bear in mind that in the suburbs this is never an easy object. Part of the difficulty is due to its large apparent size and low surface brightness, which makes the total magnitude misleading. (Remember the magnitude given refers to total brightness of the entire entity.) As one of the nearest spiral galaxies to us, it will easily fill the field of view at lowest power, and probably exceed it.

Figure 7.4.
NGC 598 (M33): drawing: narrowband filter.

Additionally, M33 is a coarse face-on spiral, this being to our disadvantage with image intensifiers. I have had the best results using a narrowband filter to reduce skyglow, once I had some idea of the nature of what I was looking for. With patience and reasonably good seeing, you will begin to suspect the S-shape structure that is the hallmark of viewing this object. Towards the tip of one of these arms lies the immense glowing gas cloud (NGC 604) that should be quite apparent as a small smudge; for orientation it is the best object to sight first. Look for the familiar features of star patterns in the field to gain orientation, and slowly but surely you will become aware of the galaxy itself. The knotty spiral arms may at first just seem to be arrangements of stars, but you will eventually sense there is some additional luminescence to them and will be able to trace out the two main spiral arms, along with the nucleus. The more transparent the skies the better, and it is especially with this object that aperture becomes an increasingly important factor.

NGC 650 (M 76) Little Dumbbell Nebula

This irregular planetary (Figure 7.5), magnitude 11, 140" × 70", with a 16.5 magnitude central star, derives its name from the casually observed similarities between it and the much larger Dumbbell Nebula (M27) in Vulpecula. Although it is often described as

Figure 7.5.
NGC 650 (M 76): drawing: narrowband filter.

one of the least conspicuous of the Messier objects, it is still quite easy to spot from our city lairs, and with a narrowband filter it becomes quite a striking sight. Time exposures reveal a complex structure with loops, altogether different to its larger namesake.

NGC 869/884 The Sword Handle Double Cluster

Although the Sword Handle Double Cluster (Figure 7.6) certainly qualifies as a supreme binocular object, it is so striking and successful even with large apertures in all locations that it can certainly justify its position on this list. Use the lowest power available in an effort to fit as much as possible of both clusters in the field of view at once. It is an amazing sight, and makes wonderful direct viewing with many star colors apparent, even without any kind of high-tech assistance. In the days when I had my $12\frac{1}{2}$-inch reflector, this was one of my favorite deep space objects, and in many ways it is most impressive at these moderate apertures, with the lower powers they allow, together with significantly higher light grasp than their smaller cousins. Intensification will bring out more stars, particularly since such a significant proportion of them are red supergiants. However, the video frames and my

Figure 7.6. Video frames: intensifier **a** The Sword Handle, NGC 869/884; **b** intensifier.

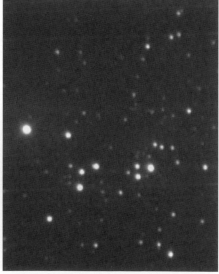

a   b

lowest magnification still do not do the subjects full justice, and they cannot both be seen together within the intensifier's relatively limited field of view.

NGC 1068 (M77)

I didn't expect much when I first surveyed this face-on and relatively compact, coarse galaxy (Figure 7.7). It did not have any particularly striking features, or so I thought. To my delight, with intensification I found a sight that clearly approximated the images I already knew. These observatory photographs reveal a prominent and mottled core, with faint and short spiral arms. On one side of the galaxy there is a prominent dark indentation toward the nucleus that appears long and notch-like in character, a feature of one of its three main spiral arms; there it was, clear as day in the intensifier view, even with slight traces of other spiral arms, also apparent on the video image with careful examination.

NGC 1097

Surprisingly, an inordinately difficult object to image on video, NGC 1097's live visual impression is far better represented by the drawing in Figure 7.8.

Observatory images show NGC 1097 to be a stunning and unusual barred galaxy. In the suburbs, at the very least, it is a great example of why it is so important to

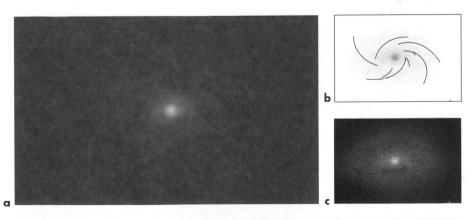

Figure 7.7. NGC 1068 (M77): **a** video frame: intensifier; **b** guide to the galaxy's structure; **c** drawing: intensifier.

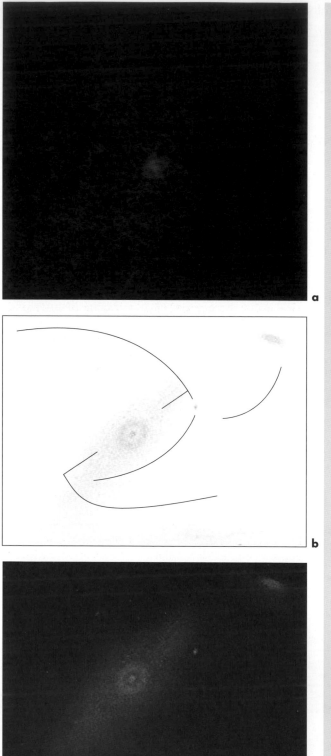

Figure 7.8.
NGC 1097: **a** video
frame: intensifier;
b guide: galaxy's
barred structure;
c drawing: intensifier.

observe with prior knowledge of a subject's photographic appearance. Without it, one would be left confused as to what was in the field of view, and possibly even unimpressed. However, in a direct intensified view, it is quite easy to recognize the major features of the galaxy, which seems to feature a "bulls eye" structured core. Towards the tip of one of its spiral arms lies a great mass of nebulosity (a companion dwarf) that will be immediately apparent in your viewing. Shreds of evidence for the existence of spiral arms become obvious now, but only once you know what you are seeking. For Northern Hemisphere observers, its low placement in the sky and limited brightness will necessitate waiting for the best of our less-than-optimal conditions, if we are to view these features successfully.

IC 418 The Spirograph Nebula

I wasn't expecting much when I first sought out this target; in suburban locations we are used to disappointments! With its description in *Burnham's* being not exactly impressive (very bright oval, 11 mag. prominent central star in nebulous disk, diam. 14″ × 11″), I thought it probably would not be a suitable object. The views from the Hubble Space Telescope certainly justified at least checking it out, so with that in mind I turned my low power eyepiece (with light filter) toward it in order to center it in the field of view. There it was, small, blue and very present to be sure, but nothing to get excited about. I couldn't have been more pleased once I switched to the image intensifier coupled to a 3× Barlow lens. The nebula, although small, is quite bright and detailed, with the dark annulus surrounding the prominent central star clear (Figure 7.9). Though the "spirograph" patterns are still far too fine to be revealed, a very pleasing visual impression of the whole object comes across, even in poor sky conditions. As is the case with some other objects, the video images fall short of the live view.

NGC 1952 (M1) The Crab Nebula

The famous Crab Nebula (Figure 7.10) is another celestial sight likely to disappoint in the suburbs.

a

b

Figure 7.9. IC 418:
a video frame:
intensifier; **b** drawing:
intensifier.

Years ago, I was seldom able even to locate it, but patience and persistence will be found to be a virtue with this object, and on unusually clear nights it may astound you. In such conditions, and using a narrow-band filter, the resemblance to well-known photographs is unmistakable, and I have been able to discern surprising amounts of the fine outer tendrils that the "Crab" is famous for. It can be seen somewhat more vaguely with intensification, but interestingly, much more successfully via video monitor than with direct viewing; the video frame here seems quite remarkable to me, especially since the famous neutron star at its center is quite evident. However, because of its low inherent brightness, it remains a challenge to be easily certain of its exact outline, all the more so when trying to draw it. Pick your night carefully, and be patient.

NGC 1976 (M42) The Great Nebula in Orion

A sight more likely to impress any viewer probably does not exist, and nothing quite prepares you for the brilliance and the complex visual spectacle with which M42 greets us (Figure 7.11). One of the grandest deep space sights we have, it has been extensively described, drawn, and photographed over the years. (It was the subject in the nineteenth century of the first astrophoto ever taken.) Without belaboring the wealth of information probably already familiar to you on this crown jewel of Orion's treasure trove, we can nevertheless discuss some approaches you might take to viewing it.

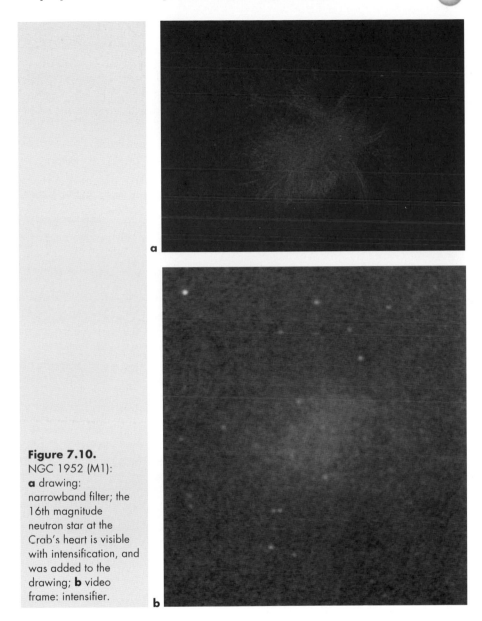

Figure 7.10.
NGC 1952 (M1):
a drawing:
narrowband filter; the
16th magnitude
neutron star at the
Crab's heart is visible
with intensification, and
was added to the
drawing; **b** video
frame: intensifier.

Overall, with or without a light filter, it will be a
magnificent sight, its twisting, swirling structure,
as well as its subtleties defying both drawing and
video image. You might try some relatively high
magnifications with no filter at all; the extensive
swirling shapes will be quite apparent with moderate
apertures. With a light filter, however, the full extent of
it will become yet more obvious, although the blue col-

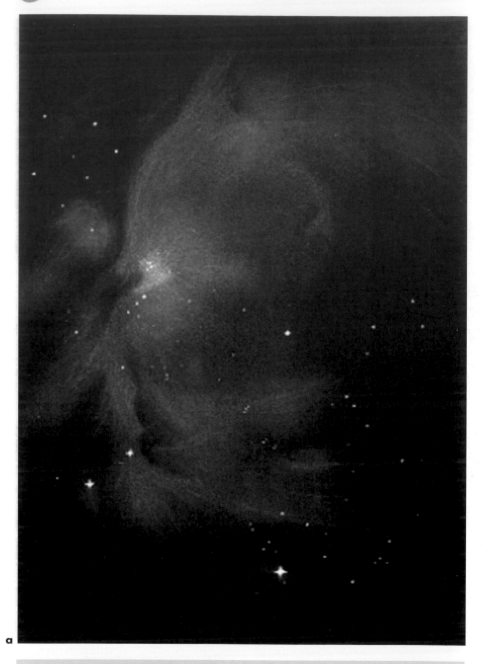

Figure 7.11. a NGC 1976 (M42): drawing: narrowband filter. Outside of time-exposed images and processing, there is really no way to adequately represent on the page the overwhelming real time magnificence and grand spectacle of this star factory; you will return to M42 many times, only to be freshly enthralled on each occasion.

b

Figure 7.11b. M 42: Wide angle video frame. View of entire region: intensifier.

oration from the filter itself will be noticeable with an object of such a great visual magnitude.

One would normally never think to use an image intensifier on such an obvious subject. However, doing so will again be a revelation, as the light emitting from the main body of it (and to a lesser degree, its extensions), is well matched to these devices (Figure 7.12). The greatest beneficiary is the Huygenian Region (at its center), which even with higher powers will blaze forth as bright as day (Figure 7.12c), the vicinity around the Trapezium revealing many components. Dozens of faint stars jump into visibility throughout the region, and twisting gaseous shapes contrast starkly with any views you may have had before; you may safely anticipate results much like familiar observatory photographs, or even some of the best amateur CCD images you may have seen.

NGC 2261 Hubble's "Variable Fan" Nebula

The variable nature of this fan-shaped nebula (Figure 7.13) was discovered by Edwin Hubble, and

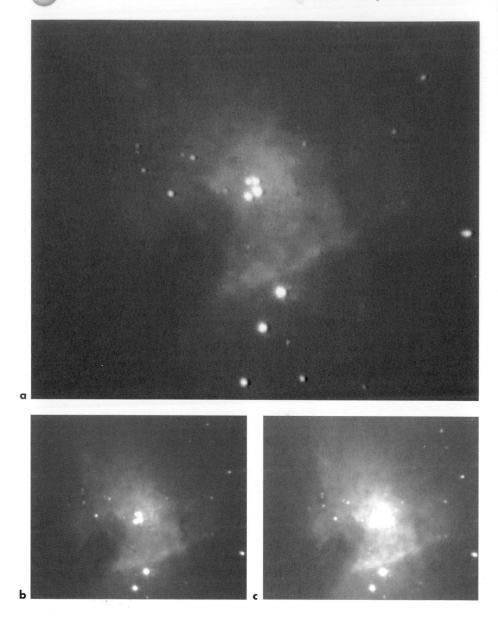

a

b   c

remained one of his favorite objects. To our good fortune, at least visually, it has a relatively high surface brightness relative to its size. Illuminated by a variable star-like object at its tip, the nebula resembles a comet in the telescope; the actual illuminating star is presumably substantially smaller than the bright blob that shows so well under intensification. Apparently, the

Figure 7.12a–c.
See caption opposite.

Figure 7.12.
a–c M42 Huygenian Region. Successive video frames of increasing saturation reveal more of the nebula, but at the expense of resolution of the Trapezium: intensifier; **d** M42 Huygenian Region drawing: intensifier/actual brightness reduced.

nebula's suitability to intensification is because of the dominant red and infrared light emission of the primary star. The much discussed variations in the nebula itself are presumably not due to true changes, since the vast dimensions of the object would preclude physical alterations occurring at such speed, something that would need to be in the order of the speed of light. You will find that its famous variations, such as have

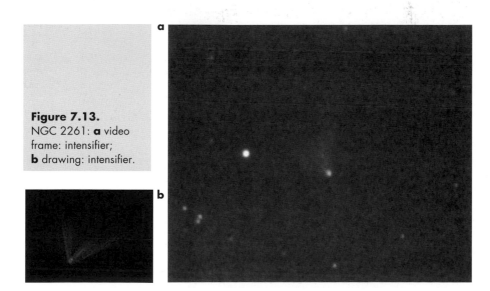

Figure 7.13.
NGC 2261: **a** video frame: intensifier; **b** drawing: intensifier.

been shown in photographic views over the years, are impossible to observe visually, especially from the city. However, it is a delight to see so readily the essence of such a well-known object of such historical importance. (Note its setting in a perfect half-circle of stars, revealed well below it in the video image.)

NGC 2359 Thor's Helmet

With its name stemming from the resemblance to the helmet of legend, this was formerly considered primarily a photographic object only. With the advent of light filters it has taken on an increased interest for the real time observer. I have had very good results, even in the typically hazy skies of my surroundings, with two approaches, and the combination of both of them provide the material for my drawing in Figure 7.14. First, with a light filter, you will begin to see the rectangular main mass of the nebula itself, along with one of the projections. As of this writing I have not been able to discern the additional lesser projections, but can foresee an occasion with the best possible skies (such as they are in suburbia) when that might happen; one can always be excused for hoping for better conditions!

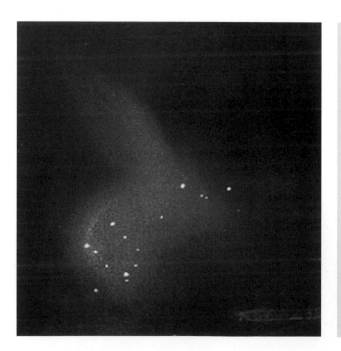

Figure 7.14.
NGC 2359: drawing: narrowband filter combined with intensifier.

The brighter embedded stars are clear, and it is easy from this to figure out the nebula's orientation. Switching to the image intensifier, the nebula virtually disappears, but the many additional stars associated with it jump into sight. By combining both views, I was able to make the drawing. All in all, a very comprehensive way to view this object.

NGC 2392 The Eskimo Nebula

I had naturally expected the best results with this famous planetary to be with the image intensifier, which does reveal much of the story: the central star, an irregular inner disk with some detail (the "face"), a dark ring surrounding it and a well-formed, mottled ring encircling the whole (the "hood"). These all do show quite clearly. Surprisingly, after trying to view it with light filters I was actually unimpressed, so I abandoned such aids and turned the telescope on it *au naturale*. With high magnification, which it took very well, much about this object remained brightly discernable, though with different emphasis and resolution of the "Eskimo" portrait as a whole from that when viewed with the I₃ (Figure 7.15).

Figure 7.15.
NGC 2392: **a** video frame: intensifier; **b** drawing: no filter/eyepiece only.

NGC 2440

This unusual planetary consists of a complex, irregular structure with bright luminous areas (Figure 7.16). In most suburban viewing situations, it would make a poor choice for viewing. Turning my image intensifier

a   b

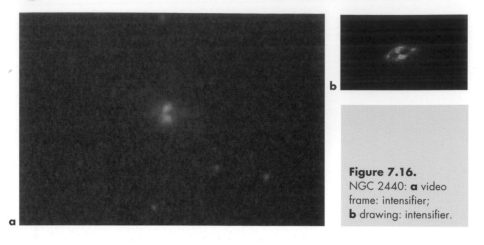

Figure 7.16.
NGC 2440: **a** video
frame: intensifier;
b drawing: intensifier.

on it is a different story. The elongated shape of the
central region, with its pointed ends, is quite apparent,
along with the bright lobe-like regions that show up so
well in images from space. It is a very striking form to
view live, utterly unlike the other planetaries so readily
available to us, but its scale is much smaller than the
grand portraits you may be familiar with. In some ways
it resembles a casual view of Saturn through a small
scope. The 16th magnitude central star, placed between
the two largest and brightest lobes, is problematic at
best. Even though my combined equipment theoretically should be able to detect it from my observing site,
I cannot lay claim to yet having seen it.

NGC 2683

This almost edge-on galaxy is one of the more unexpected delights of our survey. Normally seen at low
power in suburban surroundings as no more than an
elongated smudge, it becomes a beautiful sight when
viewed with intensification, quite sizeable, bright and
well-defined in the field of view (Figure 7.17). It is a
good example of a galaxy that responds readily to such
viewing, owing to the red and infrared spectra of light
that such edge-on galaxies present to us. In the vicinity
of NGC 2683 are scattered some foreground stars (of
our own galaxy) which frame it and make the view all
the more pleasing.

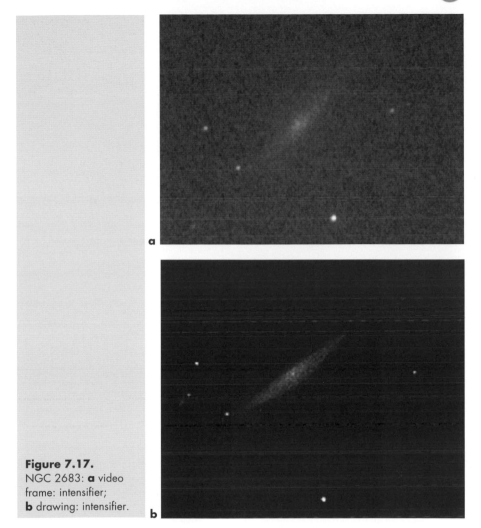

Figure 7.17.
NGC 2683: **a** video frame: intensifier; **b** drawing: intensifier.

NGC 3034 (M82)

One of the most sensational views of any galaxy I have had from my backyard location (Figure 7.18). All my memorable views of it have been through the intensifier, though I recommend that you choose the more transparent nights to experience the same degree of impact I describe here. On first glance with this device, there was such a wealth of bright detail greeting my eye that my jaw actually dropped; the illustrations here simply do not do it justice. This galaxy was so like its well-known portraits that I was stunned: extensive mottling, and complex dark veins throughout. Being so

a

b

Figure 7.18.
NGC 3034 (M82):
a video frame:
intensifier; **b** drawing:
intensifier.

bright, it requires very little adjustment of the eye. Then there is the sheer size it presents us in the eyepiece; it will fill the field of view. Strangely enough, its better-known companion M81 (close by), although known as the brighter of the two, is little more than a blurry fuzz in the eyepiece. No suggestion of its glories become evident in real time. Not so M82! You should plan on spending many sessions on this object, as it is one of our real time treasures; you will be enthralled and astounded each time, and see more detail with successive viewings.

NGC 3115 The Spindle Galaxy

NGC 3115 is a very symmetrical appearing galaxy that greatly resembles a spindle; normally faint to us from our locations, it is conspicuously revealed in the intensifier's view, and its celebrated elongated shape (also like a cigar) is immediately obvious (Figure 7.19).

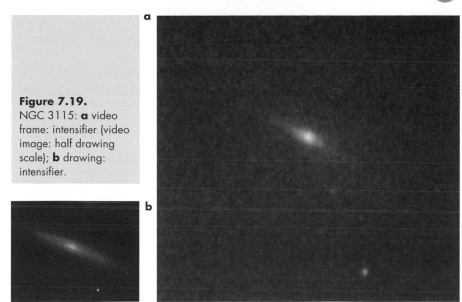

Figure 7.19.
NGC 3115: **a** video frame: intensifier (video image: half drawing scale); **b** drawing: intensifier.

As an elongated E7/SO elliptical, or transitional spiral galaxy, there are no spiral arms or other defining features, but it is a bright, prime example of its type and easy to see.

NGC 3132 Eight-Burst Nebula

Unfortunately for us in the Northern Hemisphere, this bright, large and distinctive planetary lies low in the south. However, it is possible when well placed (nearest the meridian) to catch a reasonably satisfactory view of it (Figure 7.20), all the more remarkable when we must look for it through the murkiest concentrations of atmosphere in city confines. In appearance and size, it closely resembles its better-known counterpart, the Ring Nebula M57 in Lyra, except that it is actually brighter, though less complex in ring structure (at least what is visible in real time). With the central star being very prominent and bright, instead of the challenge presented by that in M57, there will be no doubt as to its observability for any viewer. Interestingly enough, it is not this star, but a very faint companion that supplies the radiation that produces the fluorescence. Its name is derived from photographic exposures, which reveal superimposed rings at different planes.

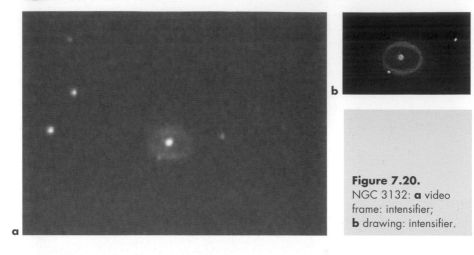

Figure 7.20.
NGC 3132: **a** video frame: intensifier; **b** drawing: intensifier.

NGC 3190 and NGC 3193

These galaxies are two parts of a well-known trio, although the third member, NGC 3187, a small barred galaxy, is difficult to detect under our suburban conditions (I have glimpsed it at times). NGC 3190 is seen edge-on and amazingly on decent nights (Figure 7.21) with intensification and sufficient aperture, it is possible to see a tiny dust lane running along its central region, somewhat offset to one side of the nucleus! Visually, it contrasts well with the elliptical galaxy NGC 3193, framed nearby in the field amongst a background of stars of varying brightnesses. Because of the relatively small field of view that my own telescope's minimum power imposes, this gathering is for me the most pleasing of real time galaxy groupings in the same field of view, available in the suburbs. Different apertures may well produce other devotees, and the brighter and wider-separated sights of the pairs of M81 and M82, and M65 and M66 may impress you more. However, it would be hard to find better contrast in galactic structures than we see in these.

An exciting moment for the author was in his unexpected "discovery" of a supernova in galaxy NGC 3190 on the night of 6 April 2002 (Figure 7.21c, NGC 3190, upper right corner). The 14th magnitude stellar explosion had in fact already been sighted, reported and logged more than a month earlier by two independent observers, but the excitement of such an unwitting "find" in real time viewing cannot be underestimated. The use of image intensification certainly was the key

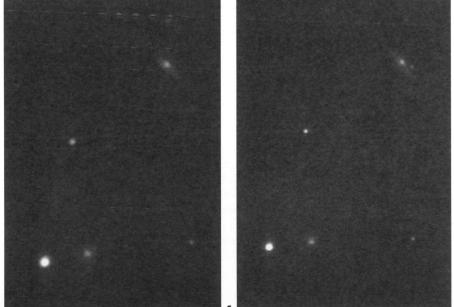

Figure 7.21. NGC 3190 and NGC 3193: **a** drawing: intensifier. Video frames: intensifier: **b** 16 December 2001, **c** 6 April 2002.

to this sighting; without it, such an event would likely go by unnoticed from suburban locations in such a diminutive subject. Never more than is the case here, the nature of video imaging often makes the finer details in the live view of some galaxies challenging to record effectively; a truer expectation is better had by examining the drawing (Figure 7.21a) and then the video image.

NGC 3242 The Ghost of Jupiter

If ever there was something in outer space that stares back at you it is this amazing planetary (Figure 7.22). It is hard to get a really good idea of this without an image intensifier, the use of which will resolve it as a circle enclosing the eye-shape itself, with its central star resembling the pupil! My first intensified view startled me, with its clear and bright resolution, not to mention the strong suggestion of a human eye. The image certainly does appear like the well-known photographs showing its more human form. Using a light filter instead of the image intensifier, it is easy to see why it is also known as the Ghost of Jupiter, since aside from the similar shape and size, the more transparent presence of the nebula does suggest something of a spiritual apparition of the giant planet. There are other planetaries with something of this elliptical shape as a hallmark, but none impose so strongly a living personality as does this one.

a

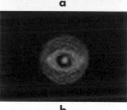

b

Figure 7.22.
NGC 3242, the Ghost of Jupiter, also known as the Eye Nebula: **a** video frame: intensifier; **b** drawing: intensifier.

NGC 4111

This edge-on galaxy presents us with one of the most unlikely suburban surprises. An elliptical (E7 or S0) lenticular galaxy at magnitude 11.6 and size 3.4' × 0.8', it hardly seems a candidate for spectacular viewing, but intensified, it reveals itself prominently, albeit relatively small, as typical of its category, with two bright stars well above its axis to complete the view (Figure 7.23). It is a strangely beautiful object, suggesting to me a flying saucer in rendezvous with other spacecraft, the striking setting demanding repeated viewing.

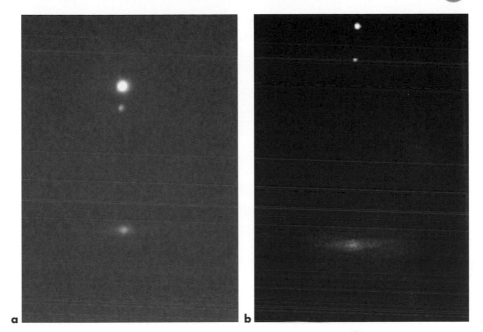

a   b

NGC 4374 (M84)/4406 (M86) Region of the Center of the Virgo Cluster

Individually, the ellipticals M84 and M86 are not exceptional sights for us. What is important about them is their close proximity to the center of the enormous swarm of galaxies known as the Virgo Cluster (Figure 7.24). I did not feel it helpful to include any real time drawings or other illustrations, because in the suburbs you will be unlikely to be able to see more than one or two galaxies at a time in the field of view, even with moderate apertures. These galaxies individually are not especially striking: studied as a group, though, it is an entirely different matter. Begin with the central region for a thorough exploration; whether this is a success will depend on your individual circumstances and equipment, though I should add that with smaller apertures the area will not likely reveal many bright galaxies. Be sure to reference *Burnham's*.

NGC 4565

The most famous of the edge-on galaxies, NGC 4565 will not disappoint you with the spectacle it provides,

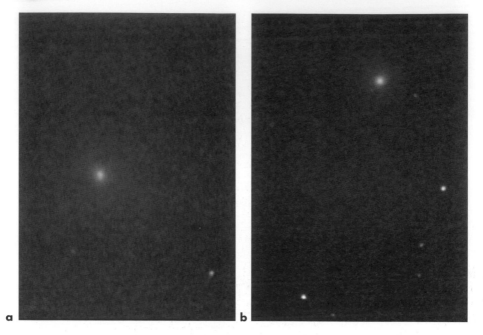

a   b

as long as you make an allowance for its less-than-expected total brightness. Not normally considered an impressive suburban object, with care and time, this galaxy presents a view that closely resembles the images we are used to seeing in textbooks; it is one of the grand prizes of deep space viewing (Figure 7.25). Be sure to use averted vision to see all that there is to see. All in all, it is a simply astounding sight, the famous dust lane seeming to trail off into space itself. This is also one of the relatively few galaxies that gives us such ideal apparent dimensions in the field of view, the size and general impression being so suggestive of its celebrated observatory portraits.

Figure 7.24.
NGC 4374 (M84) and NGC 4406 (M86): Adjoining video frames: intensifier: **a** M86, **b** M84. These bright ellipticals, at one end of what is known as "Markarian's Chain", with the arc of galaxies forming the most prominent members of the cluster, are a good starting point for telescopic explorations in the region.

NGC 4594 (M104) The Sombrero Galaxy

Here is another of the greatest splendors for the real time suburban observer (Figure 7.26). Even without a light filter something is immediately obvious about this remarkable and bright galaxy. The image intensifier and ever-greater apertures make clear what is so different about it: the dark band sweeping so dramatically across the edge-on view, along with the sombrero-

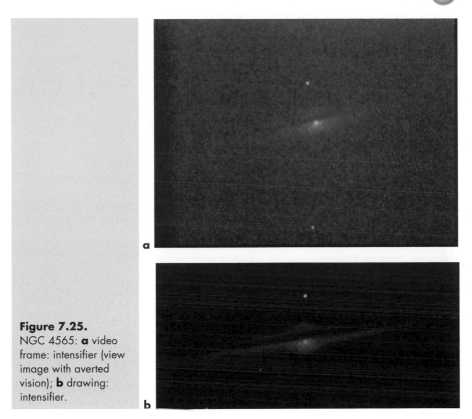

Figure 7.25.
NGC 4565: **a** video frame: intensifier (view image with averted vision); **b** drawing: intensifier.

shaped glow rising from its midst. If you look at it with indirect vision, you are likely to see the maximum detail and shape, and you will be even more aware of the encircling band, which seems almost darker than the surrounding space. Not much imagination is needed here, as the Sombrero lives up to its name so readily and comes very close to observatory images. It is also particularly satisfying in the perfect dimensions that it presents us in the field of view.

NGC 4736 (M94)

To those who always say what they see is too faint, this galaxy, even without intensification, should do the trick (Figure 7.27). Viewed with the intensifier, its stunning bright core sets it apart from any other, with the exception of M31. It is one of the brightest such objects we can view from our suburban locations. However, in real time it looks more like a very bright elliptical than

Figure 7.26. NGC 4594 (M104): **a** video frame: intensifier (averted vision reveals the true outline); **b** drawing: intensifier.

the face-on compact spiral that it is. Its concentrated form makes it difficult to resolve anywhere near its nucleus with photography as well, and it is one of the most compressed spirals of those in our cosmic neighborhood. Look for the two bright stars (Alpha and Beta Canum) that form a right-angle triangle with it in the corner.

NGC 4826 (M64) The Black Eye Galaxy

This is yet another prize for the suburban viewer (Figure 7.28). Some aperture is needed to show the famous "black eye" clearly, even with image intensification, but it is there plainly enough with a reasonable combination of favorable circumstances. Although the live view shows no traces of the galaxy's spiral form, obvious in time exposures, this is one more galactic sight that will easily live up to expectations. The "black eye", that great cloud of obscuring

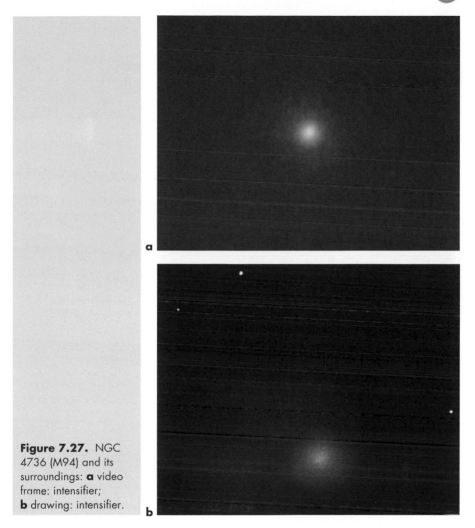

Figure 7.27. NGC 4736 (M94) and its surroundings: **a** video frame: intensifier; **b** drawing: intensifier.

dust, stands out clearly, within the shape of the readily seen galaxy, at least in my own viewing experience.

NGC 5128 Centaurus A

NGC 5128 is one of my favorite galaxies to view, even though it is amongst the worst placed for Northern Hemisphere observers, lying in a barely accessible part of the sky for my own location, above the unobscured ocean horizon. It would be quite difficult for us to contend with, with our dubious city air, but becomes a

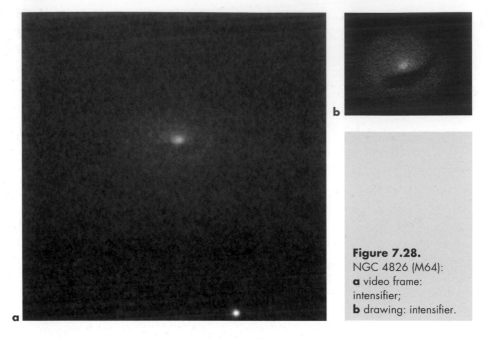

Figure 7.28.
NGC 4826 (M64):
a video frame:
intensifier;
b drawing: intensifier.

prime object with an image intensifier (Figure 7.29). I tend to favor carefully chosen times during winter months for viewing it (usually in the early hours of the morning). This is because of the problem I usually experience with haze later in the year when it rises at a decent hour. Look for it when it is near the meridian, as then it will be highest in the sky; check out carefully when it will cross that imaginary line. For latitudes further south of this (34°), it is a much easier object, but sadly, observers too much to the north may not be able to access it.

The actual classification of this "peculiar" galaxy is still being debated, but it appears to be elliptical or even SO, with a not quite explained massive dark equatorial band. With intensification, it provides a stunning and haunting view, with the dark band (and its bright central division) running across it. Since the core region on many photographs is often overexposed, the actual size of the galaxy is far greater than is at first obvious. Getting oriented even after careful prior referencing may still be difficult. You have to realize that you are seeing only the brightest portion of the galaxy, and the central band is so wide that you may mistake the bright inner part for one of the two hemispheres of this galaxy. It is only when you look further at various photographic images, and begin to appreciate the huge

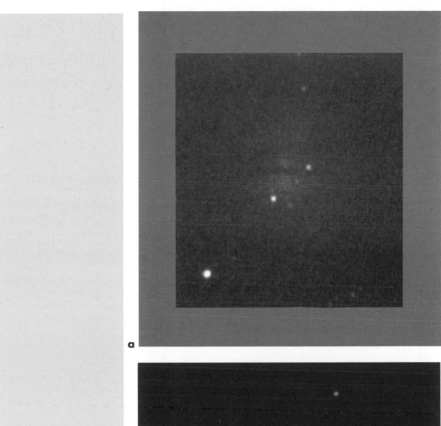

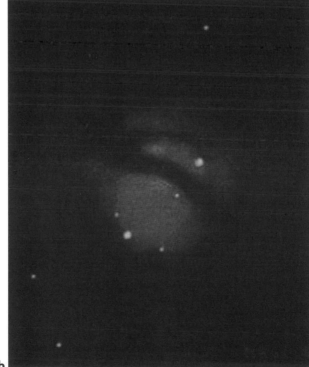

Figure 7.29.
NGC 5128: **a** video frame: intensifier (careful examination will reveal some remarkable detail in this image); **b** drawing: intensifier.

scale that it presents to us, that it becomes clear what you are actually seeing.

NGC 5139 Omega Centauri

I include this object here (Figure 7.30), the king of all globulars, knowing well that it may lie too far to the south for most Northern Hemisphere observers. It is so bright that it remains an easy mark for city observers, assuming that it is possible to access it above the horizon. Even in Southern California it is not an easy object, lying yet further south of NGC 5128 (previous paragraph), and I am always grateful to have the occasional opportunity to see it; doing so is a special event. Whatever extreme efforts you may need to take to gain even a glimpse of this mightiest of globulars will be highly rewarding, and you may be completely unprepared for the tremendous spectacle. Looking over the ocean on a good night to near the horizon from my location, the great cluster presents itself as a faintly glowing spot to the unaided eye.

The best times to view it are when viewing galaxy NGC 5128, as it lies nearby, and poses similar problems to access. Ungodly hours in the morning, during winter, may again often be our best bet. Even through the thick atmospheric layering (and in my case murk), if you can turn your telescope on it, what you see will astound you. Omega Centauri is a huge ball of stars, far

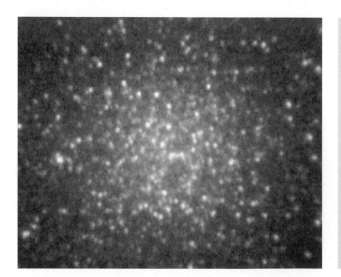

Figure 7.30.
NGC 5139: video frame: intensifier.

larger than anything else we have to compare it to; certainly the description of "swarm" applies here! Use of my image intensifier resolves the whole ball into a mass of glowing points that fills the entire field of view. Look for the darker regions within the whole, visible in the image here. You might try to view it from a hilltop where the horizon is well below your vantage point, and there are no obscuring terrestrial objects. If you do succeed, this magnificent "swarm" will not disappoint you.

Figure 7.31.
NGC 5236 (M83):
a video frame:
intensifier; **b** guide;
c drawing: intensifier.
One of the brightest
galaxies available to
us, M83 nevertheless is
much vaguer than you
might expect and poses
some observing
challenges, which is
why the guide sketch is
included; it should
make clear the general
placement of the spiral
arms relative to what
we see.

NGC 5236 (M83)

I must confess that the viewing possibilities suggested by knowing the magnitude and size of this spectacular barred galaxy prepared me for something a lot more imposing than the actual observing experience turned out to be. Even with the best tools I have to offer, M83 remains somewhat elusive in the suburbs. However, it is still well worth viewing, and I have found that with the image intensifier it gives a decent account of itself (Figure 7.31). I wish more of the total extent of it was visible, but we have to be satisfied with a fairly bright view of the central hub, revealing several small and

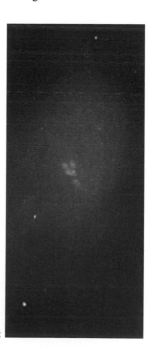

a   b   c

unequally bright lobes centrally placed amongst a vaguely luminous larger oval object, the main body of the galaxy. The views of the central lobes actually compare quite favorably with observatory views of the galaxy hub, which naturally reveal better resolved clumps of brightness in the same positions.

NGC 5272 (M3)

As one of the finest and most beautiful globulars in the sky, M3 is quite compact and striking telescopically (Figure 7.32). The scale and brightness it presents us ranks it alongside M13 and M5 as luminaries in the Northern skies. It is an obvious sight for us in our dubious surroundings. Under image intensification M3's nucleus appears quite brilliant but lopsided, throwing into greater contrast the small patches of dark obscuring matter against the brighter background. It is currently unknown if these regions, similarly seen in many other globulars, are actually present in the structure itself or non-luminous matter projected against the background of stars. Visually, they certainly appear to be integral to the cluster itself. As with some of the other bright and highly impressive globulars, you may also be struck by what appear to be streamers of stars, spirally radiating outwards from the hub of the cluster.

Figure 7.32.
NGC 5272 (M3):
video frame: intensifier.

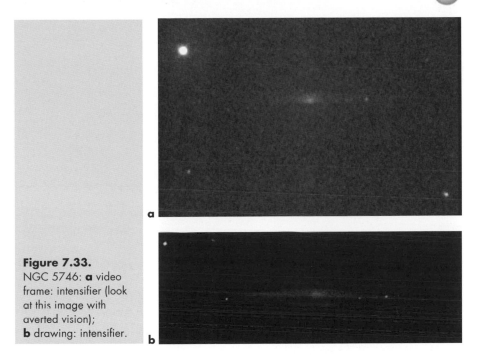

Figure 7.33.
NGC 5746: **a** video
frame: intensifier (look
at this image with
averted vision);
b drawing: intensifier.

NGC 5746

Another wonderful edge-on galaxy that shows up well
under intensification in suburban skies (Figure 7.33),
despite its relative faintness (magnitude 11.7).
Apparently a barred galaxy (the presence of a
"peanut"-shaped core on time-exposed images indicates this), the prominent equatorial dust lane shows
itself fairly easily with sufficient apertures. The bright
starry condensations at the outer edges are true bright
"knots" of the galaxy itself, and contribute greatly to
the interest of this subject. Don't expect NGC 5746 to
equal the more famous NGC 4565, however, which
remains the prize amongst this group. However, this
galaxy presents a large and striking enough object to
rank as premium galaxy viewing.

NGC 5866

In *Burnham's*, the magnitude of this wonderful and
interesting galaxy is given as 11.1 – not the most

Figure 7.34. NGC 5866:
a video frame: intensifier;
b drawing: intensifier. The third
star framing this galaxy is
surprisingly less obvious on the
video frame than it appears to
the eye in the live view, although
some faint stars closer to the
galaxy (between the two
brightest foreground stars) may
be just detectable, and which I
did not note at the time of my
drawing. This again underlines
the differences of the two types
of image and the necessity to
remain objective in weighing
expectations.

encouraging numbers for our suburban location! However, with the total brightness enhanced by its small size, I couldn't have been more delighted (Figure 7.34). It immediately ranked amongst the best deep sky sights I know. Easily located in my low power eyepiece, it appeared to be up to a full magnitude brighter, and promised great things under image intensification. Framed by three stars, we see a cigar-shaped mass, shaped not unlike observatory portraits, and best of all, the very narrow central dust lane jumps out at first glance. Quite a prize! I would have thought that this fine band would not be possible to see in real time even with 18-inch aperture, but you can probably still expect it to show in substantially smaller telescopes. It is a unique and memorable sight amongst all those available to us.

NGC 5904 (M5)

One of the finest globulars known, M5 is, however, little more than half the apparent size of M13. As with

Figure 7.35.
NGC 5904 (M5):
video frame: intensifier.

all clusters, in our suburban locations it will benefit greatly from image intensification, and when conditions merit, it can be impressive even with direct viewing, with or without a light filter (Figure 7.35). It possesses one of the most densely packed cores of such objects, and as with many others, is not perfectly round in shape. The intensified appearance also reveals other aspects that are generally not discussed. It is clear to see in the video image above that M5 has a narrow but prominent dark lane dissecting one third of the core from the remainder, and other branches of obscuring matter. The core is lop-sided in brightness on the opposite side of the dark lane, and it shows extensive trails of stars radiating outwards from the cluster. These have suggested circular formations around the cluster to many observers. M5's (also strikingly M92's) capacity to suggest spiral form is even greater than with most other globulars. After Lord Rosse discovered spiral formation in galaxies, he tended to see everything with some degree of this type of structure. However, some globulars such as M5 do indeed seem to offer some partial validation of his findings.

NGC 6205 (M13)

This legendary globular (Figure 7.36) is the finest in the Northern Hemisphere (M22 is a Southern Hemisphere cluster). An interesting quirk surrounding this sight

Figure 7.36.
NGC 6205 (M13):
video frame: intensifier.

concerns the so-called "propeller lanes", some of the most famous of their type. First described by Lord Rosse in the nineteenth century, they seemed to have apparently "vanished" until around 1980, when again they began to be reported by various observers. Strangely enough, they can easily be seen on numerous photographs from before that date, but somehow were overlooked. This became clear to me after reading about the "rediscovery" and then almost immediately afterwards saw these lanes clearly showing on a photograph from the 1960s in Dr. Henry Paul's book *Telescopes for Skygazing*. Additionally, they are also extensively discussed by Burnham in his milestone work (*Burnham's Celestial Handbook*) published in the 1970s, but compiled in the 1960s. They sometimes show on other photographs, depending on exposure times, and this may be the cause of their "disappearance," since lengthy exposures tend to fog the lanes.

Look for three equally spaced dark lanes, converging to a central point, appearing well towards one side of the cluster (Figure 7.37). Rosse's drawing shows them as more centrally placed, and this may have also contributed to their "disappearance", especially since they don't immediately "jump out" at the observer. Photographs show a narrower range in brightness across the cluster than in real time viewing, so be prepared to look into regions that are less brightly illum-

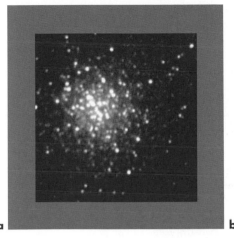

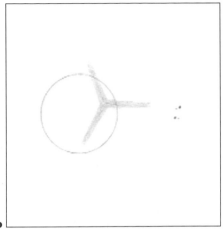

a   b

Figure 7.37.
NGC 6205 (M13):
a video frame:
intensifier; **b** guide.

inated. Also be aware that the scale of the real time image at magnifications over 100× will be larger than in most photographs, and so the lanes are broader than you may be anticipating. Plot them against recognizable star patterns toward the edges of the cluster on a photograph beforehand. (Especially, you might look for what appears to be a "double–double" star group, actually a "double–triple", toward the outside of the cluster. The outward-pointing vein of the "propeller" points toward this.) Initially, the lanes may not be readily obvious, but once you have them in sight they stand out unmistakably; they are especially elusive in the suburbs without image intensification.

Although they do not show prominently in Figure 7.36, they are clearly revealed in Figure 7.37a because of its smaller size, and also the greater contrast obtained with it. (Actually, this image brightness is not unlike the overall appearance on a superior night.) The simple guide sketch included, with the same orientation, should make the lanes' detection foolproof, as well as the position of the guide stars I mention. The degree of sky transparency, your telescope's aperture, the amount of light pollution and the performance of your view-enhancing devices are all, of course, factors. While an image intensifier makes the lanes easily visible, it even more drastically increases the range of light amplification on individual stars, so the "propellers" seem at first to hide amongst other potential lanes of different star brightnesses. You may also be successful in viewing them with a light filter, or even simply a high magnification to emphasize darker areas, once you know what to look for.

The sight of these famous symmetrical lanes is as fascinating as it is tempting to be definitive about their nature, here or in other deep space structures. We still don't know exactly what they are.

NGC 6266 (M62)

A small but bright globular, M62 is distinguished by its irregular stellar distribution and shape (Figure 7.38). Through the eyepiece it is prominent and no less striking than many larger globulars. You will notice numerous fine star points, the majority being of 14th. and 15th. magnitudes, concentrated in a small area, coming to a concentrated blaze off-center. The effect of this is quite different to some other globulars, such as M92, which often do not show nearly the same concentrated populations of stars, because of their closer proximity to us.

NGC 6302 The Butterfly (or "Bug") Nebula

This unusual planetary has been described as appearing like a flattened figure eight (Figure 7.39). I can certainly see where that description might come from, although I have never been able to see the com-

Figure 7.38.
NGC 6266 (M62): video frame: intensifier.

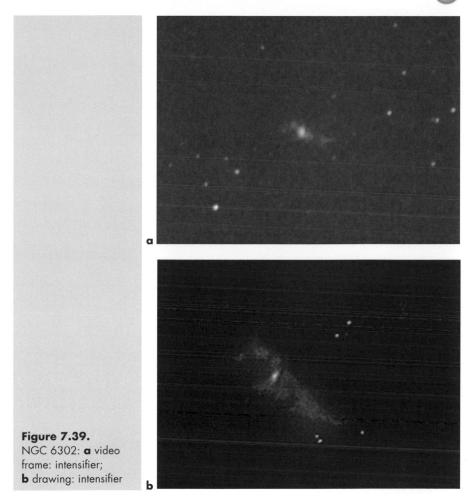

a

b

Figure 7.39.
NGC 6302: **a** video
frame: intensifier;
b drawing: intensifier

pletion of either loop. Observatory images resolve its
true form. There are no loops, but it is indeed a
bipolar nebula that has two shells of gas at each side
of the central star, in some ways casually resembling
loops, but more like a butterfly than anything else.
The nebula immediately reveals itself with just a narrowband filter, but it presents us the greatest subtlety
of detail with image intensification, the image
showing very keenly the brightly formed areas, as
well as the central star. All kinds of knots, and
regions of varying brightness can be seen, making for
interesting study, and it is unlike any other object so
readily available to us.

NGC 6341 (M92)

One of the smaller globular clusters to make our primary list, M92 is hardly unimposing (Figure 7.40). If the grand and famous M13 had not been so close by (9º), there is little doubt that its smaller neighbor would have been better known and recognized as one of the more striking examples of its kind. It certainly deserves considerable attention, particularly as we scour the skies for worthwhile "suburban" objects. Flying high in the sky for most Northern Hemisphere observers during summer months, it is in a prime location for viewing through the thinnest air, and the superior seeing that results from this. Intensified viewing shows a quite uneven but bright nucleus, presumably due once again to the presence of true dark matter in the cluster itself, or somewhere in the foreground. As in other notable examples, stars seem to spiral around the circumference of the cluster. This to me is the best example of this phenomenon, a visually compelling case for spiral formation of some kind. Lord Rosse would be pleased. Hopefully, one day the issue of whether this feature of many globulars is merely an illusion, or indeed real, will be settled once and for all.

NGC 6369 Little Gem

This planetary is included here more as a demonstration and test for the effectiveness of your equipment than it is a promise of a spectacular sight. It is, never-

Figure 7.40.
NGC 6341 (M92): video frame: intensifier.

Figure 7.41.
NGC 6369: video frame: intensifier (look carefully at the center of the image!).

theless, a remarkable object. If you can make it out against suburban bright skies with sufficient aperture, its perfect and finely shaped ring and central star make a textbook example of such nebula formations. With fair skies, image intensifier and 18-inch aperture, this relatively faint planetary shows clearly, and amazingly enough, the 16th magnitude star at its center is quite readily revealed (Figure 7.41). Such a remarkable feat would be out of the question in suburban conditions without a 40–50-inch telescope, which speaks volumes again as to the value of the Collins I_3 image intensifier, pulling this image out of the competing skyglow.

NGC 6402 (M14)

As the faintest globular on our primary list, M14 is a good example of an extremely rich cluster of relatively evenly matched stars, mostly of 15th and 16th magnitudes. Increasing apertures will reveal this object to its best advantage: a tightly packed mass of minute stellar points, like fine dust or sand, grouped tightly together to form an exquisite and delicate celestial ornament (Figure 7.42).

NGC 6514 (M20) The Trifid Nebula

One of the great showpieces of the sky, the Trifid Nebula (Figure 7.43) is one of the very few real time

Figure 7.42.
NGC 6402 (M14):
video frame: intensifier.

suburban nebulae that exhibit color in the eyepiece
(with sufficient aperture). With my Ultrablock filter, its
predominant red color is quite apparent at first glance,
and even a hint of the blue region on the north side is
evident (although that may be due partially to the narrowband filter emphasizing this part of the spectrum).
The three primary lanes of dark matter crossing it are
very obvious, and they intersect near the center of the
nebula. In one of the central "corners" caused by this
intersection lies the famous multiple star which excites
the luminosity of the nebula itself. Depending on the
viewing conditions, the magnification used, and the
aperture of your telescope, it may only be resolvable,
though, as a double star. Surprisingly, you will not find
intensified viewing of M20 particularly satisfactory, as
the nebula will likely fade into obscurity; however, in
my telescope, the multiple star resolves into no less
than five components, two of them lying only 2.2″
apart. The crowning glory of the Trifid is its setting in a
sea of countless stars, and it will surely assume a place
at or near the top of your list.

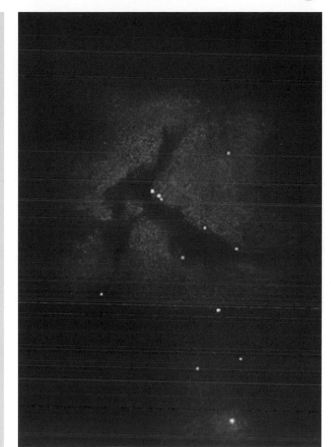

Figure 7.43.
NGC 6514 (M20):
drawing: narrowband
filter.

NGC 6523 (M8) The Lagoon Nebula

The magnificent Lagoon Nebula has few parallels in the entire sky (Figure 7.44). Easily accessible to amateur observers, it readily lends itself to viewing from the confines of suburbia, particularly so with the use of light filters. It appears bright and studded with stars, the dark lane (the "lagoon" itself) running through its mid-section, and clearly resolved with the slightest of effort. Within its boundaries lies the dazzling open star cluster NGC 6530, further lending near magical qualities to the nebula. The best real time views will probably be with narrowband filter, the beauty of form being plain to see. With the star cluster glowing within, it is one of the most memorable and mystically beautiful

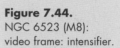

Figure 7.44.
NGC 6523 (M8):
video frame: intensifier.

sights in the entire sky. The results are also quite pleasing with real time imaging using image intensification; by cross-referencing the images here against the drawing (shown at the equivalent of a lower power to include a greater area) it becomes quite clear what parts of the nebula are revealed on the video image.

Highly luminous emission regions of many nebulae such as this are also strong sources of radio waves, and the nebulae that contain them typically respond favorably to all viewing approaches. Look for the subtle dark lanes in this region, which will be plainly apparent in live viewing with a light filter. They are visible only subtly on the video image, but are still discernable under careful examination.

The richly populated star field is reminiscent of the Trifid Nebula region, and the bright nebulosity surrounding it is striking even in limited apertures. Spectacular as the Lagoon Nebula is in real time though, you should not expect to see the nebula's finer features, such as the dark "globules" visible on photographs.

The drawing in Figure 7.45 gives a good low power visual representation of the Lagoon Nebula's appearance with a light pollution filter. You will notice the different visual emphasis, especially in the Hourglass Region, under image intensification (Figure 7.46). With a narrowband filter, this feature is visible as a brightening in the middle of the nebula, becoming strongly emphasized and starkly revealed under image intensification, its shape clearly defined. (It is approximately 30″ in size.) It is interesting that while this par-

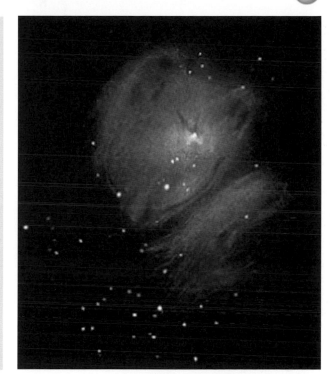

Figure 7.45.
Drawing of NGC 6523
(M8).

ticularly luminous region is highly enhanced with intensification, much of the remainder of the nebula becomes faint. In the high-power close-up in Figure 7.46, the quality of detail of the region is quite remarkable.

Figure 7.46.
NGC 6523 (M8): The "Hourglass Region" close-up video frame: intensifier

NGC 6543 The Cat's Eye Nebula

As a planetary, the Cats Eye Nebula is a small but very striking object; with intensified viewing it is particularly so, both in brightness and the complexity of structure it begins to let us see (Figure 7.47). It has some unusual features: projections spinning out of two sides seem to hint at its having a twisted or helical structure, which is confirmed by observatory images. With or without narrowband filters, it is an easy object in the amateur's telescope, with very high surface brightness, although nothing of its structure and the central star

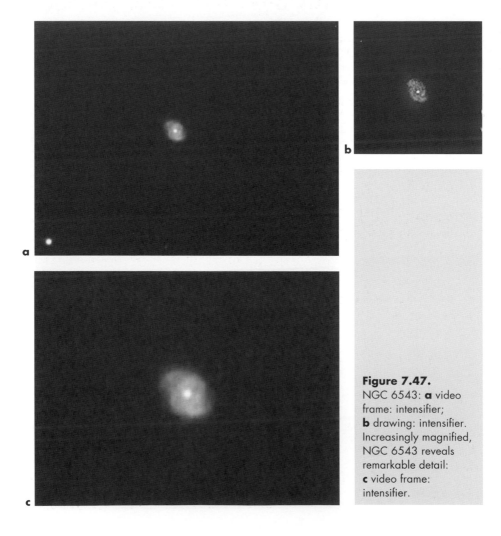

Figure 7.47.
NGC 6543: **a** video frame: intensifier; **b** drawing: intensifier. Increasingly magnified, NGC 6543 reveals remarkable detail: **c** video frame: intensifier.

are likely to be seen with the apertures commonly available. Under image intensification, a revelation occurs. The central star, with its surrounding annulus, becomes quite evident; abundant mottling and even "spokes" seem to radiate from the star itself. In the steadiest conditions, it may be possible to discern even more of the helical and twisting structure itself. It is amongst the finest planetaries available to us in the suburbs in real time, and will withstand even considerable light and atmospheric pollution. Figure 7.47 show the typical intensified real time appearance at moderate powers.

NGC 6611 (M16) Star Cluster with the Eagle Nebula

One of the most photographed sights in the sky, M16 (which refers to the well-populated open cluster throughout it), is situated in a field of remarkable nebulosity (Figure 7.48). The famous "eagle" itself – the dark nebula superimposed on the luminosity – is a difficult object in the suburbs. I have glimpsed it with a narrowband filter on transparent nights, although it remains invisible with the image intensifier from my location. The luminous regions are not to be counted amongst the brightest available to us; earlier observers were actually unaware of the nebula's existence. Though the cluster is easy with the slightest optical aid, just how much of the fantastic diffuse and dark nebulae that you will actually see in direct viewing depends on your specific circumstances and the light grasp of your telescope (Figure 7.49). Nevertheless, M16 should provide considerable viewing interest and challenge, demanding repeated viewing; it is indeed very satisfying to detect something of the "eagle" if your own circumstances allow it.

NGC 6618 (M17) The Omega or Swan Nebula

The Omega Nebula is one of the finest gaseous condensations available to the real time amateur observer (Figure 7.50). It is so bright and prominent that it is unlikely to disappoint you within city limits, and in

Figure 7.48. NGC 6611 (M16), the Eagle Nebula with Cluster M16, as imaged by W.J. Collins in Denver, Colorado, using a 7-inch Astro-Physics apochromatic refractor, coupled to a Collins Electro Optics I₃ Piece and Canon D30 digital camera, in a 6-second exposure. The hallmark eagle feature is clear on this image, as it appears to loom out of the background toward the cluster itself. This approximates the real time view with a decent aperture from Collin's own suburban surroundings. In comparison to the author's, they feature generally much lower humidity with far greater transparency. In such skies it is not inconceivable that such a view as this, in real time, would indeed be possible under image intensification, with larger apertures such as the author utilizes, since such conditions are more receptive to intensified viewing; quite a prospect!

conditions that would obliterate most other deep space objects. It is one of Sagittarius' seemingly disproportionate claims to many of the most remarkable objects in the visible universe. Resembling a vast swan (by which it is also known), it responds favorably to light filters and image intensifiers. Depending on your telescope's aperture and sky conditions, M17 will reveal increasing complexity and filamentary structure. When viewed either with or without a light filter, much of these complex yellowish swirls and filaments will still show in a poor sky.

Figure 7.49. NGC 6611 (M16). On this video frame, on a larger scale than Figure 7.48, it is possible to make out definite traces of nebulosity, and possibly even some hints of part of the dark nebulae superimposed. (Compare it directly with Figure 7.48.) There are numerous columns of this dark nebulosity possible to see under your best conditions. In my situation they are best revealed with a narrowband filter.

NGC 6656 (M22)

This spectacular Southern Hemisphere cluster, M22, is the largest and grandest globular easily accessible to those of us living in the Northern Hemisphere. In terms of size and brightness, it ranks third of all globulars, only after Omega Centauri and 47 Tucanae. In the Northern Hemisphere, only M13 itself ranks close to it, although the appearance of that cluster is quite different, being more circular, denser and even. M22's appearance is not unlike a somewhat smaller, more open Omega Centauri. Stunning, even with unaided telescopic viewing, M22 comes even more into its own with image intensifier, and dazzlingly so (Figure 7.51). Through this device it is truly magnificent, filling the field of view with a multitude of bright stellar points, featuring many highly prominent members amongst them.

a

b

Figure 7.50.
NGC 6618 (M17):
a video frame:
intensifier; **b** the
"Swan" M17, riding in
a sea of stars, sails the
skies! drawing:
narrowband filter. M17
even responds
favorably to the use of
no enhancing devices
at all, and will provide
striking views even in
the presence of some
phases of the Moon.

NGC 6705 (M11) The Wild Duck Cluster

It is quite easy to imagine similarly the flock of ducks that Admiral Smythe described upon sighting M11, which also happens to be one of the most spectacular of the open clusters, rivaling even some of the globulars (Figure 7.52). It is certainly much more than a "binocular object", and makes for beautiful viewing. The shallow "V" shape line (upper right) that leads the "flock of ducks" is evident enough, followed by the pointed shape of the flock itself, although on time

Figure 7.51.
NGC 6656 (M22):
video frame: intensifier.

exposures these features become far less obvious and tend to disappear. Real time observing preserves the shape and general impression of M11, and it is those features that make the cluster all the more interesting

Figure 7.52
NGC 6705 (M11):
video frame: intensifier.

to me. You will probably find that intensified viewing of most clusters is usually preferable because of the resolutions of fainter stars, and overall image brightness. There is never any doubt about the nature of what we see, although it does come at the expense of real colors and the curious three-dimensional illusion of natural viewing.

NGC 6720 (M57) The Ring Nebula

Probably the most celebrated of planetaries, this famous sight is immediately apparent in the eyepiece without filters or even image intensification (Figure 7.53). However, use of a narrowband filter emphasizes its brilliance and contrast against the sky, and shows its brightly glowing blue color. Once we make the switch to the intensifier, while the overall brightness and contrast are not necessarily enhanced, subtle shadings, gradations and wisps appear in the ring itself. Then there is the greatest bonus of all: on a reasonably good night, the central star, so easily missed by many observers with even very large telescopes, is at last plain to see with little effort using moderate-to-larger amateur scopes! I can even see it quite unmistakably on the monitor when using CCD video. This in spite of its light falling in the blue part of the spectrum, such wavelengths not being favored by image intensifiers. Within the central "void", I don't see the bands often described and photographed, but the area is not as dark as the surrounding black of space. This nebula is a guaranteed crowd pleaser, and will easily justify frequent visits of your own.

NGC 6826 The Blinking Nebula

This prominent planetary derives its name from a technique used by many observers to view the central star of such systems. When looking directly at the main ring of the nebula itself, hence diverting one's attention away from the star, a tiny central point of light will suddenly jump into view, courtesy of the effects of averted vision. Looking directly at the star, it will seem

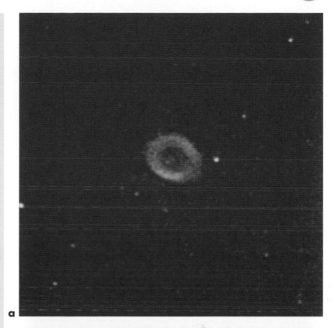

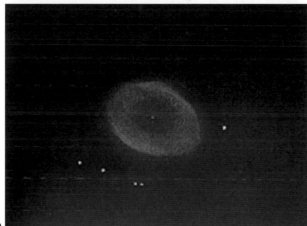

Figure 7.53. NGC 6720 (M57): **a** In this real time video image (video frame: intensifier), the central star can just be seen, along with its visual companion. This illuminating central star, a famous visual challenge, is possibly a variable, since it is observably inconsistent from night to night. Some of this is due to conditions and the sensitive nature of faint stars in brighter surroundings, but this does not necessarily explain the particular phenomenon in this instance. **b** Drawing: intensifier.

to disappear and the nebula becomes more prominent instead. With increased aperture, the so-called blinking effect is reduced. With my 18-inch all aspects of the nebula are easy all at the same time, even without a light filter, and particularly so at high magnifications (Figure 7.54). Intensification provides a different perspective still, with the central star becoming even more prominent in a central blaze.

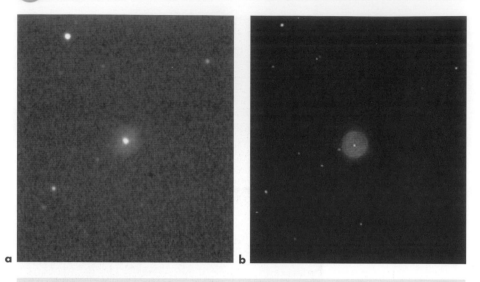

a   b

Figure 7.54. NGC 6826: **a** video frame: intensifier; **b** drawing: unaided.

NGC 6853 (M27) The Dumbbell Nebula

With or without a narrowband filter, the Dumbbell Nebula makes its presence known immediately, even in our compromised skies. This is not one of those sights that you suspect you can see! It appears as a bright white egg-shape in the field of view, quite luminous and striking (Figure 7.55). Sufficient aperture will resolve its famous outline easily, and shadings of brightness become obvious. The subtle extensions on either side may be seen with larger scopes, and give the nebula a somewhat different overall outline, much more elongated (as in my drawing in Figure 7.55a). The illuminating star is a hard catch, though in good viewing conditions, you may possibly succeed. It just shows on the video frame (Figure 7.55b) at the center of the "bar". An image intensifier works in an interesting way; the nebula fades substantially, but many stars within and around it, unseen without intensification, become obvious to the point that perceptions of the internal structure of the nebula itself seem to be influenced by the stellar outlines (Figure 7.55). This, of course, adds a new degree of visual understanding. As a deep sky sight, it is another "sure thing" for visitors, as one of the few celestial objects that they may actually have heard of and can see so clearly.

Figure 7.55. NGC 6853 (M27):
a drawing: narrowband filter. The disparity between the visual and video images is affected by the eye's different sensitivities to frequencies of light, opposed to that of image intensifiers and video cameras. Although viewing the nebula will probably be to greatest effect with a narrowband filter, I believe the appearance and general outline of the central "bar" is even affected by ambient light from the linear arrange-ment of the stars, those essentially unseen when viewing with a light filter but clearly showing in the video image.
b Video frame: intensifier (look care-fully!).

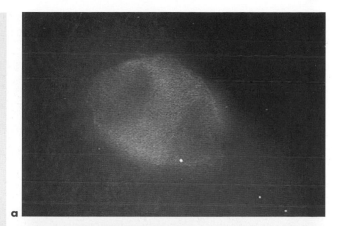
a

b

NGC 6960/6992 The Veil Nebula

One of the most celebrated nebulae, the Veil Nebula used to be one that routinely disappointed its would-be viewers. Thanks to the magic of light filters, it has become something of an almost easy mark on good nights, even with the sky conditions that form the focus of this book. The two NGC numbers refer to the two primary parts of the nebula, on each side of the area it covers (Figures 7.56 and 7.57). (A third component of the "Veil" is far less prominent, and a more central part of it, but it is unlikely to be detectable.)

Certainly this whole interstellar gas structure reveals evidence of being the result of some gigantic cosmic explosion many thousands of years ago. A narrowband

filter will likely aid greatly in revealing a sight much like well-known portraits, although significantly less bright. Don't expect to see color, however; its hue is the pale white/gray characteristic of most live deep space viewing. Depending on your scope's aperture, though, you may well be surprised with just how much of it you can clearly resolve, with much delicacy of the "filamentary" features revealed. Image intensification will not be successful with the "Veil", even though the light is of the emission variety; the specific wavelengths and total brightness make all the difference.

Figure 7.56.
NGC 6992: narrowband filter.

Figure 7.57.
NGC 6960: narrowband filter.

Figure 7.58.
NGC 7008: drawing:
narrowband filter.

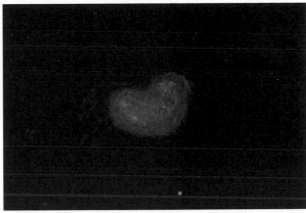

NGC 7008

This curious, rather faint little planetary nebula occupies a place on this list because it seems almost unique with its heart-shaped form (Figure 7.58). Careful examination may also reveal some mottling. Do not expect to be dazzled by its presence, however, and it will not respond favorably to image intensification. For whatever reason, I cannot help being led back to it for another look.

NGC 7009 The Saturn Nebula

Named the Saturn Nebula by Lord Rosse, this planetary probably first revealed its true nature to this nineteenth-century figure with his newly constructed 72-inch reflector. Fine CCD or observatory views will at once make the reason for its name clear, as it has on each side the most extraordinary projecting ansae (Figure 7.59). These give the nebula a very Saturn-like appearance. From our unfavorable vantage point, however, the extremities of these are likely to remain unseen under the best of conditions and with the largest amateur scopes. I have, on occasion, seen hints of these extremities, particularly by moving the object in the field of view to allow indirect vision to reveal their presence. The rest of the object is a splendid sight regardless, with the central star, two surrounding ovals and projecting irregularities prominently displayed for us under intensification. It is interesting to note that

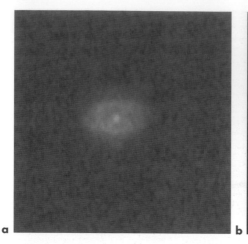

a　　　　　　　　　　　　　b

Rosse's great telescope was probably unable to reveal the central star or the darker central region, both quite easy with image intensification and a quarter the aperture of his great instrument.

Figure 7.59.
NGC 7009: **a** video frame: intensifier; **b** drawing: intensifier.

NGC 7027

This tiny planetary is well worth the trouble to seek out and study, even from our suburban locations. It appears as a bright star-like core with two illuminated lobes on one side (Figure 7.60). Faint extensions to these lobes are also just visible. You will also see narrow separations between all three of these main components. It is necessary to utilize higher magnifications to see it well, and certainly you will need them to discern any detail, because it is so diminutive. The video image here was produced with a Barlow lens, as with many of the other smaller and brighter subjects, and shows NGC 7027 approximating the way it appears with powers between 200× and 300×.

NGC 7078 (M15)

Truly one of the jewels of the skies, M15 is a relatively compact globular, and slightly oblate in shape (Figure 7.61). Through the intensified eyepiece, this oblate shape is visually quite noticeable in the shape of the core, a tightly knit mass ablaze with stellar points. Also within this area, there are some dark lanes that

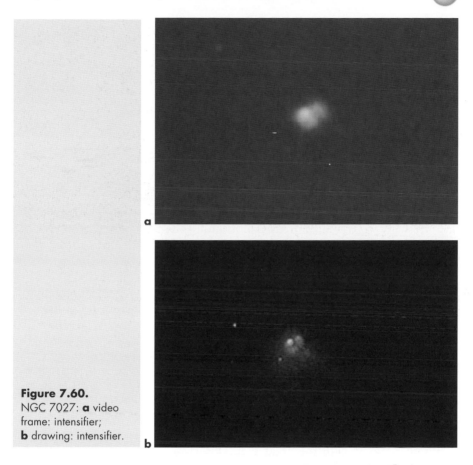

Figure 7.60.
NGC 7027: **a** video frame: intensifier; **b** drawing: intensifier.

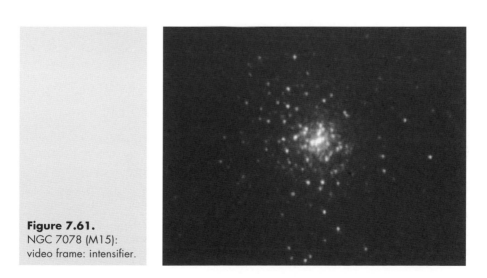

Figure 7.61.
NGC 7078 (M15): video frame: intensifier.

show quite readily. The cluster is notable for its great concentration of stars.

NGC 7089 (M2)

M2 is another of the more notable globulars, bright but small only because of its great distance. This makes its stellar population much fainter and densely packed than is the case with the closer clusters. Like M15, it is somewhat oblate in appearance, and under intensified viewing it reveals a wonderful, evenly distributed compact array of fine, bright star points (Figure 7.62). It is unusually well revealed in my Collins unit, and in many ways more magically than most globulars. Near the northwest corner is a prominent dark lane, which shows well on the video image, and it can easily be seen under good conditions with image intensification. Of all the globular star clusters, M2 is amongst the most satisfying and beautiful to view.

NGC 7662

One of the brighter and more striking planetaries, NGC 7662 will always remind me of the great E.E. Barnard, who drew it so impressively at the beginning of the twentieth century, with the great Lick Observatory 36-inch refractor. Seen under intensification in significantly less apertures, it reveals much the same

Figure 7.62.
NGC 7089 (M2):
video frame: intensifier

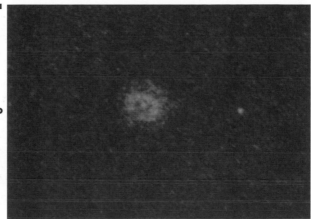

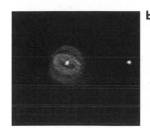

Figure 7.63.
NGC 7662: **a** video frame: intensifier; **b** drawing: intensifier.

sight that Barnard saw, with the central star being an easy mark within the very well-defined annulus, even though this star is a blue dwarf (Figure 7.63). What appears casually like a double ring is not so much unusual as it is beautifully formed and detailed. The nebula is noticeably oval in shape, with a flattening on one of the longer sides.

This, then, completes our survey of the most successfully viewed objects from our suburban environs, based on my own perspective and viewing experiences. With all that conspires against us, it is surprising to me just how many sights there are in suburbia that are still so accessible and rewarding in so many ways. Happily, in Chapter 8 you will find that the possibilities go on, and many of the sights listed there may even find their way onto your own preferred list.

Chapter 8

Second Viewing Catalog

The purpose of this chapter is to provide you with a guide to other deep space objects that may likely be successfully viewed in real time from the suburbs; for me, it is not enough that the potential is just sufficient to be able to make them out against the background sky. None of them were included in my best list, of course. Some were close contenders. Many though, fell short when viewed in my own circumstances, and by a considerable degree. However, all are worth your perusal, as it is possible that your own particular situation will make at least some of them become part of your own personal best list. They are all worth the effort to seek out, as everything presented here was considered very carefully. Nevertheless, you may find additional objects that could be included if your own situation differs greatly from mine (including your geographic location in latitude). Further north or south, you may find others. However, in weighing the widest possible range of criteria for inclusion, I have tried to present the most complete list possible covering a wide range of locations. Much of this list applies equally to Northern and Southern Hemisphere locations, and includes certain relevant information on each object. The listing gives NGC catalog number (or other catalog number if it applies), magnitude (where possible), and certain information in many instances concerning outstanding features, as well as size, co-ordinates and constellation.

From our suburban locations, lengthy and detailed descriptions of the objects would be largely redundant, as they are typically less successful to view than those of the primary list. However, it is possible that any of these on the secondary list may well provide you with some impressive results. Clearly though, preparing the eye for complex detail it is not likely to see would be misleading. There are some video frames included of some of the most striking objects, recorded as in Chapter 7. As for the best way to view them, with filter, intensifier, or other, I leave this to your own experimentation. You will find that, armed with the experience gained from the objects detailed in the previous chapter, the best approach will usually follow your inclinations. I believe you will enjoy these additional explorations and am confident that this catalog will guide you well; maybe you can even expand it. Heaven knows, there is no shortage of sights to look for.

Open Clusters

The following list includes many of the so-called binocular objects. Open clusters tend to be less spectacular than globular clusters in the more highly magnified fields of telescopes because of their wide and relatively sparse stellar distribution. You may agree or disagree whether some should have been counted amongst the best celestial objects to view from suburban locations in Chapter 7. You will certainly be able to see most of them, regardless.

NGC 1039 (M34)	magnitude 6, angular size 20', celestial coordinates (02420n4247), constellation **Perseus**.
NGC 1912 (M38)	magnitude 7.2, angular size 20', celestial coordinates (05287n3550), constellation **Auriga**.
NGC 1931	magnitude 12.1, angular size 3', celestial coordinates (05314n3415), constellation **Auriga**. With "comet-like" nebulosity.
NGC 1960 (M36)	magnitude 6.8, angular size 12', celestial coordinates (05361n3408), constellation **Auriga**.
NGC 1981	magnitude 5.4, angular size 25', celestial coordinates (05352s0426), constellation **Orion**. Near M42.
NGC 2099	magnitude 6.4, angular size 20', celes-

(M37)	tial coordinates (05524n3233), constellation **Auriga**. Densely populated.
NGC 2158	magnitude 11, angular size 4′, celestial coordinates (06075n2406), constellation **Gemini**. Challenging but interesting arrowhead-shaped cluster adjacent to M35.
NGC 2168 (M35)	magnitude 5.5, angular size 30′, celestial coordinates (06089n2420), constellation **Gemini**. With open cluster NGC 2158.
NGC 2237/ 2244	magnitude 5.5, angular size 40′, celestial coordinates (06324n0452), constellation **Monoceros**. **Rosette Complex:** cluster and nebula (Figure 8.1). See NGC 2237 notes in the Emission, Reflection and Dark Nebulae listing later in this chapter. The center of the cluster sits right in the heart of the rosette itself, in the central "hole". If you use a very low-power eyepiece, the complete outline will fit within the field of view, and you should be able to see something of the famous shape of the nebula as well.
NGC 2264	celestial coordinates (06411n0953), constellation **Monoceros**. **Christmas Tree Cluster**; Large and spread out.

Figure 8.1. Center of the Rosette Cluster, NGC 2244: video frame: intensifier.

Much nebulosity present, including the dark **Cone Nebula**, though it is unlikely you will see more than the famous Christmas tree outline of the cluster. See also the Emission, Reflection and Dark Nebulae section.

NGC 2287 (M41)
magnitude 6, angular size 30′, celestial coordinates (06470s2044), constellation **Canis Major**.

NGC 2323 (M50)
magnitude 6, angular size 10′, celestial coordinates (07032s0820), constellation **Monoceros**.

NGC 2420
magnitude 9, angular size 7′, celestial coordinates (07385n2134), constellation **Gemini**. Remote; resolved with I_3.

NGC 2422 (M47)
magnitude 5, angular size 20′, celestial coordinates (07366s1430), constellation **Puppis**.

NGC 2437 (M46)
magnitude 8, angular size 25′, celestial coordinates (07418s1449), constellation **Puppis**. With planetary NGC 2438; use NBF.

NGC 2447 (M93)
magnitude 7, angular size 18′, celestial coordinates (07446s2352), constellation **Puppis**.

NGC 2477
magnitude 7, angular size 25′, celestial coordinates (07523s3833), constellation **Puppis**. Densely populated (Figure 8.2).

Figure 8.2.
NGC 2477: video frame: intensifier.

NGC 2539	magnitude 6.5, angular size 20′, celestial coordinates (08107s1250), constellation **Puppis**.
NGC 2682 (M67)	magnitude 7, angular size 15′, celestial coordinates (08504n1149), constellation **Cancer**.
NGC 6231	magnitude 6, angular size 15′, celestial coordinates (16540s4148), constellation **Scorpius**.
NGC 6405 (M6)	magnitude 6, angular size 25′, celestial coordinates (17401s3213), constellation **Scorpius**. Fine cluster.
NGC 6475 (M7)	magnitude 5, angular size 60′, celestial coordinates (17539s3449), constellation **Scorpius**. Fine cluster.
NGC 6494 (M23)	magnitude 7, angular size 25′, celestial coordinates (17568s1901), constellation **Sagittarius**. Use lowest power; 9–13 mag. stars.
NGC 6520	magnitude 9, angular size 5′, celestial coordinates (18034s2754), constellation **Sagittarius**. Open cluster, enclosed by M24 – Sagittarius Star Cloud. A fine cluster, good with intensified viewing. Look for dark nebula B 86 nearby, next to 7 mag. star; looking like a dark hole, it is quite apparent with INT.
NGC 6531 (M21)	magnitude 7, angular size 10′, celestial coordinates (18046s2230), constellation **Sagittarius**. In field with Trifid Nebula; see Chapter 7.
NGC 6603	magnitude 11.1, angular size 4′, celestial coordinates (18184s1825), constellation **Sagittarius**. Enclosed by M24 – Sagittarius Star Cloud.
NGC 6694 (M26)	magnitude 9.5, angular size 9′, celestial coordinates (18452s0924), constellation **Scutum**.
NGC 6838 (M71)	magnitude 9, angular size 6′, celestial coordinates (19538n1847), constellation **Sagitta**. Rich and compact: possibly a globular, although it lacks a dense core (Figure 8.3). Stars approximately 12 mag.
NGC 7654 (M52)	magnitude 7, angular size 12′, celestial coordinates (23242n6135), constellation **Cassiopeia**. Improves with aperture.

Figure 8.3.
NGC 6838 (M71):
video frame: intensifier.

NGC 7789   magnitude 10, angular size 20', celestial
            coordinates (23570n5644), constella-
            tion **Cassiopeia**. 900 + stars.

Globular Clusters

These beautiful objects will present themselves well to
you in a number of ways. Typically, they are bright
enough to show even without a light filter. However,
for resolution and ease of seeing, it is hard to beat the
image intensifier, although the cluster will seem more
one-dimensional and not reveal any star color. Viewed
in this manner, there will be no doubt as to the globulars' spectacular form, and star points will usually
resolve out of what may only have seemed like nebulosity before. Some of the best reactions from visitors
are with intensified views of globulars.

Hopefully I have covered many of the best globulars
in the previous chapter. However, you should not conclude that the second list below will not provide some
magnificent viewing, since this variety of deep space
object tends to be amongst the most impressive. There
are many other globulars as well, to be sure (all fainter
and smaller), but probably the best or at least most
interesting, outside those presented in Chapter 7, are
presented here.

NGC 288   magnitude 7.2, angular size 10', celestial
           coordinates (00528s2635), constellation
           Sculptor. Relatively sparse globular.

Figure 8.4.
NGC 6093 (M80):
video frame: intensifier.

NGC 1904 (M79)	magnitude 8.4, angular size 7.4′, celestial coordinates (05245s2433), constellation **Lepus**. Faint, resolved with INT.
NGC 2419	magnitude 11.5, angular size 2′, celestial coordinates (07381n3853), constellation **Lynx**. The most distant Milky Way globular; interesting to see. Unresolved with INT.
NGC 4590 (M68)	magnitude 8, angular size 9′, celestial coordinates (12395s2645), constellation **Hydra**.
NGC 5024 (M53)	magnitude 8, angular size 10′, celestial coordinates (13129n1810), constellation **Coma Berenices**. 1° from fainter NGC 5053.
NGC 6093 (M80)	magnitude 8, angular size 7′, celestial coordinates (16170s2259), constellation **Scorpius**. Small and bright; appears to radiate spikes of stars, mostly 14 and 15 mag. (Figure 8.4).
NGC 6121 (M4)	magnitude 7.4, angular size 20′, celestial coordinates (16236s2632), constellation **Scorpius**. Large, open and relatively sparse; known for its striking loops and chains of brighter stars; look for the bright equatorial bar of central stars (Figure 8.5).

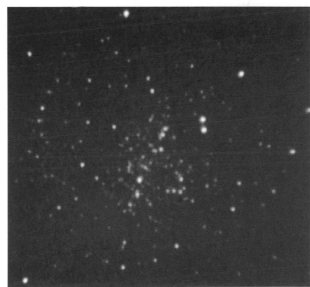

Figure 8.5.
NGC 6121 (M4):
video frame: intensifier.

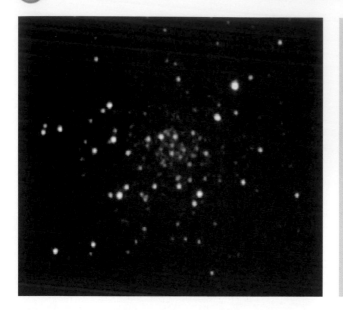

Figure 8.6.
NGC 6218 (M12):
video frame: intensifier.

NGC 6218 (M12)	magnitude 8, angular size 10′, celestial coordinates (16472s0157), constellation **Ophiuchus**. Fairly sparse (Figure 8.6).
NGC 6254 (M10)	magnitude 7, angular size 8′, celestial coordinates (16571s0406), constellation **Ophiuchus**. See Figure 8.7. Near to globular cluster NGC 6218 (M12).
NGC 6273 (M19)	magnitude 7, angular size 6′, celestial coordinates (17026s2616), constellation

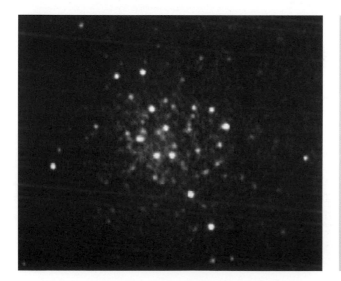

Figure 8.7.
NGC 6254 (M10):
video frame: intensifier.

Ophiuchus. Oblate; near the center of the Milky Way; faint star population.

NGC 6333 (M9) — magnitude 8, angular size 4′, celestial coordinates (17192s1831), constellation **Ophiuchus.**

NGC 6522 — magnitude 10.5, angular size 2′, (18036s3002), constellation **Sagittarius.** In field with:

NGC 6528 — magnitude 11, angular size 1′, celestial coordinates (18048s3003), constellation **Sagittarius.**

NGC 6626 (M28) — magnitude 8, angular size 6′, celestial coordinates (18245s2452), constellation **Sagittarius.** Bright and dense.

NGC 6637 (M69) — magnitude 7.5, angular size 4′, celestial coordinates (18314s3221), constellation **Sagittarius.** Appearance unimposing in smaller telescopes; resolution of 14 and 15 mag. stars needs larger apertures. Near 9 mag. star (Figure 8.8).

NGC 6681 (M70) — magnitude 8, angular size 4′, celestial coordinates (18422s3218), constellation **Sagittarius.** Uneven distribution.

NGC 6715 (M54) — magnitude 9, angular size 6′, celestial coordinates (18551s3029), constellation **Sagittarius.** Compact and bright; a remarkable globular, apparently as magnificent as Omega Centauri, although situated outside our own

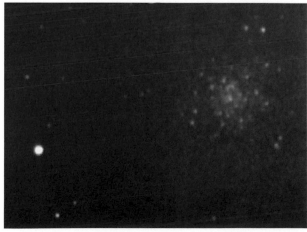

Figure 8.8.
NGC 6637 (M69): video frame: intensifier.

galaxy (Figure 8.9). Belonging to the Sagittarius Dwarf Galaxy – not to be confused with NGC 6822 – it requires larger apertures to resolve any of its stars; an image intensifier will greatly aid in this. The video image here reveals some stellar resolution.

Figure 8.9.
NGC 6715 (M54): video frame: intensifier.

NGC 6779 (M56)   magnitude 8, angular size 5′, celestial coordinates (19166n3011), constellation **Lyra**. Mostly 11–14 mag. stars. Unusual location for a globular; it is not especially bright, but is quite pleasing to view, and challenging for smaller scopes to resolve (Figure 8.10).

NGC 6809 (M55)   magnitude 7, angular size 15′, celestial coordinates (19400s3058), constellation **Sagittarius**. Large and loose; stars mostly fainter than 11 mag.

NGC 6864 (M75)   magnitude 8, angular size 3′, celestial coordinates (20061s2155), constellation **Sagittarius**. Fairly bright, compact and dense; most stars 17 mag. Partly resolved with INT.

NGC 6981   magnitude 8.6, angular size 3′, celestial coordinates (20535s1232), constellation **Aquarius**. Brightest stars 15 mag.

NGC 7099 (M30)   magnitude 8, angular size 6′, celestial coordinates (21404s2311), constellation **Capricorn**.

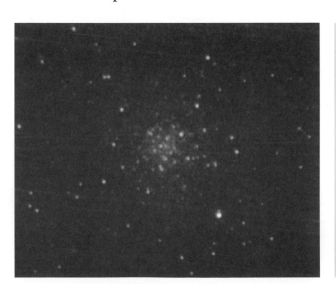

Figure 8.10.
NGC 6779 (M56): video frame: intensifier.

Galaxies

Many of these fall into the gray area of whether they should have been included at all; many others may show far better from your own location than mine.

NGC 55   magnitude 7.8, angular size 25′ × 4′, celestial coordinates (00149s3911), constellation **Sculptor**. Irr. or SBp galaxy in some ways like M82; would probably be one of my prime objects were it not for the haziness of my location during the times it is visible.

NGC 247   magnitude 10, angular size 18′ × 5′, celestial coordinates (00471s2146), constellation **Cetus**. Sc galaxy.

NGC 300   magnitude 11.3, angular size 21′ × 14′, celestial coordinates (00549s3741), constellation **Sculptor**. Sc/Sd galaxy; S-shape.

NGC 404   magnitude 11.9, angular size 1.3′ × 1.3′, celestial coordinates (01094n3543), constellation **Andromeda**. E0/S0 galaxy.

NGC 488   magnitude 11.2, angular size 3.5′ × 3′, celestial coordinates (01218n0515), constellation **Pisces**. Sb galaxy; very compact; nearly face-on.

NGC 628 (M74)   magnitude 11, angular size 9′ × 9′, celestial coordinates (01367n1547), constellation **Pisces**. Sc galaxy; face-on.

NGC 891   magnitude 12.2, angular size 12′ × 1′, celestial coordinates (02226n4221), constellation **Andromeda**. Sb galaxy; edge-on with equatorial dust lane; difficult object and fainter version of NGC 4565; see Chapter 7.

NGC 1023   magnitude 11, angular size 8.6′ × 4.2′, celestial coordinates (02404n3904), constellation **Perseus**. E7 galaxy; lens shape with satellite galaxy on eastern edge.

NGC 1232   magnitude 10.7, angular size 7′ × 6′, celestial coordinates (03098s2035), constellation **Eridanus**. Sc galaxy.

NGC 1291 — magnitude 10.2, angular size 5′ × 2′, celestial coordinates (03173s4108), constellation **Eridanus**. SB galaxy.

NGC 1300 — magnitude 11.3, angular size 6′ × 3.2′, celestial coordinates (03197s1925), constellation **Eridanus**. SB galaxy.

NGC 1316 — magnitude 10.1, angular size 3.5′ × 2.5′, celestial coordinates (03227s3712), constellation **Fornax**. SO galaxy. With NGC 1317, magnitude 12.2, angular size 0.7′ × 0.6′, SB galaxy.

NGC 1398 — magnitude 10.7, angular size 4.5′ × 3.8′, celestial coordinates (03389s2620), constellation **Fornax**. SBb galaxy.

NGC 1399 — magnitude 10.9, angular size 1.4′ × 1.4′, celestial coordinates (03385s3527), constellation **Fornax**. EO galaxy; brightest of the Fornax cluster (9 in total).

NGC 1792 — magnitude 10.7, angular size 3′ × 1′, celestial coordinates (05052s3759), constellation **Columba**. Sc galaxy.

NGC 2403 — magnitude 8.8, angular size 16′ × 10′, celestial coordinates (07369n6536), constellation **Camelopardalis**. Sc galaxy.

NGC 2613 — magnitude 10.9, angular size 6.4′ × 1.5′, celestial coordinates (08334s2258), constellation **Pyxis**. Sb galaxy; edge-on.

NGC 2841 — magnitude 10.3, angular size 6.2′ × 2′, celestial coordinates (09220n5058), constellation **Ursa Major**. Sb galaxy.

NGC 2903 — magnitude 9.7, angular size 11′ × 4.7′, celestial coordinates (09322n2130), constellation **Leo**. Sb/Sc galaxy; elongated/fair.

NGC 2964 — magnitude 11.9, angular size 2.3′ × 1.1′, celestial coordinates (09429n3151), constellation **Leo**. Sb/Sc galaxy.

NGC 2976 — magnitude 10.8, angular size 3.4′ × 1.3′, celestial coordinates (09432n6808), constellation **Ursa Major**. Sc/Sd/Irr. galaxy; mottled.

NGC 3031 (M81) — magnitude 8.9, angular size 18′ × 10′, celestial coordinates (09556n6904), constellation **Ursa Major**. Sb galaxy; one of the most beautiful spirals in the sky;

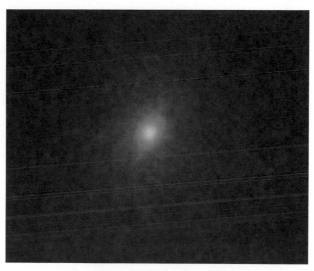

Figure 8.11.
NGC 3031 (M81):
video frame: intensifier.

spiral structure not apparent or revealed in real time (Figure 8.11). In field with NGC 3034 (M82); see Chapter 7.

NGC 3109   magnitude 11.2, angular size 11′ × 2′, celestial coordinates (10031s2609), constellation **Hydra**. Irr. galaxy.

NGC 3184   magnitude 10.5, angular size 5.5′ × 5.5′, celestial coordinates (10183n4125), constellation **Ursa Major**. Sc galaxy; face-on.

NGC 3351 (M95)   magnitude 11, angular size 4′ × 3′, celestial coordinates (10440n1142), constellation **Leo**. SBb galaxy. In field with:

NGC 3368 (M96)   magnitude 10.2, angular size 6′ × 4′, celestial coordinates (10468n1149), constellation **Leo**. Sb galaxy; fuzz.

NGC 3379 (M105)   magnitude 10.6, angular size 2.1′ × 2′, celestial coordinates (10478n1235), constellation **Leo**. E1 galaxy. In field with:

NGC 3384   magnitude 11, angular size 4′ × 2′, celestial coordinates (10483n1238), constellation **Leo**. E7/Sc galaxy.

NGC 3521   magnitude 10.2, angular size 6′× 4′, celestial coordinates (11058n0002), constellation **Leo**. Sb galaxy; right nucleus, elongated.

Figure 8.12. NGC 3623 (M65): video frame: intensifier.

Figure 8.13. NGC 3627 (M66): video frame: intensifier.

NGC 3556	magnitude 10.8, angular size 7.8′ × 1.4′, celestial coordinates (11115n5540), constellation **Ursa Major**. Sc galaxy; dust lanes and stellar nucleus; near the Owl Nebula.
NGC 3621	magnitude 10.6, angular size 5′ × 2′, celestial coordinates (11183s3249), constellation **Hydra**. Sc/Sd galaxy; bordered by stars.
NGC 3623 (M65)	magnitude 10.3, angular size 7.8′ × 1.6′, celestial coordinates (11189n1305), constellation **Leo**. Sa/Sb galaxy; elongated (Figure 8.12). In field with:
NGC 3627 (M66)	magnitude 9.7, angular size 8′ × 2.5′, celestial coordinates (11202n1259), constellation **Leo**. Sb galaxy (Figure 8.13).
NGC 3628	magnitude 10.3, angular size 12′ × 2′, celestial coordinates (11203n1336), constellation **Leo**. Sb galaxy; edge-on.
NGC 3810	magnitude 11.5, angular size 3.6′ × 2.5′, celestial coordinates (11410n1128), constellation **Leo**. Sc galaxy.

NGC 3941	magnitude 11.3, angular size 1.9' × 1.1', celestial coordinates (11529n3659), constellation **Ursa Major**. E3/SO galaxy.
NGC 3992 (M109)	magnitude 10.9, angular size 6.4' × 3.5', celestial coordinates (11576n5323), constellation **Ursa Major**. SBb galaxy.
NGC 4088	magnitude11.1, angular size 4.7' × 1.5', celestial coordinates (12056n5033), constellation **Ursa Major**. Sb/Sc galaxy; mass to one side.
NGC 4096	magnitude11.5, angular size 4.1' × 1.1', celestial coordinates (12060n4729), constellation **Ursa Major**. Sc galaxy; edge-on.
GNG 4125	magnitude11.1, angular size 2.1' × 1.1', celestial coordinates (12081n6511), constellation **Draco**. E5/SO galaxy.
NGC 4192 (M98)	magnitude 11, angular size 8.2' × 2', celestial coordinates (12138n1454), constellation **Coma Berenices**. Sb galaxy; edge-on.
NGC 4214	magnitude10.5, angular size 7' × 4.5', celestial coordinates (12156n3620), constellation **Canes Venatici**. Irr or early SB galaxy.
NGC 4216	magnitude10.9, angular size 7.2' × 1', celestial coordinates (12159n1309), constellation **Virgo**. Sb galaxy; thin edge-on with two others in field. Near the center of the Virgo Cluster.
NGC 4217	magnitude 11.9, angular size 4' × 1', celestial coordinates (12158n4706), constellation **Canes Venatici**. Sb galaxy; with dust lane.
NGC 4244	magnitude10.7, angular size 13' × 1', celestial coordinates (12175n3749), constellation **Canes Venatici**. Sb galaxy; edge-on with streak.
NGC 4254 (M99)	magnitude 10.4, angular size 4.5' × 4', celestial coordinates (12188n1425), constellation **Coma Berenices**. Sc galaxy.
NGC 4258 (M106)	magnitude 9, angular size 19.5' × 6.5', celestial coordinates (12190n4718), constellation **Canes Venatici**. Sb galaxy; fairly striking (Figure 8.14).

NGC 4303 (M61)	magnitude 10.2, angular size 5.7′ × 5.5′, celestial coordinates (12219n0428), constellation **Virgo**. Sc galaxy; face-on.
NGC 4321 (M100)	magnitude 10.4, angular size 5.2′ × 5′, celestial coordinates (122229n1547), constellation **Coma Berenices**. Sc galaxy.
NGC 4382 (M85)	magnitude 10.5, angular size 3′ × 2′, celestial coordinates (12254n1811), constellation **Coma Berenices**. E galaxy.
NGC 4449	magnitude 10.5, angular size 4.2′ × 3′, celestial coordinates (12282n4406), constellation **Canes Venatici**. Irr. galaxy.
NGC 4472 (M49)	magnitude 10.1, angular size 4′ × 3.4′, celestial coordinates (12298n0800), constellation **Virgo**. E3/E4 galaxy – one of the largest.
NGC 4486 (M87)	magnitude 10.1, angular size 3′ × 3′, celestial coordinates (12308n1224), constellation **Virgo**. E1 galaxy: giant; famous jet not visible from my location.
NGC 4490	magnitude10.1, angular size 5′ × 2′, celestial coordinates (12306n4138), constellation **Canes Venatici. Cocoon Galaxy**; Sc galaxy; pear-shape with NGC 4485, magnitude12.5, angular size 1.3′ × 0.7′, Irr. or E galaxy.
NGC 4501 (M88)	magnitude 10.5, angular size 5.7′ × 2.5′, celestial coordinates (12320n1425), constellation **Coma Berenices**. Sb galaxy.

NGC 4526	magnitude10.7, angular size 4′ × 1′, celestial coordinates (12340n0742), constellation **Virgo**. E7/SO galaxy; edge-on, between two 7 mag. stars.
NGC 4535	magnitude 10.7, angular size 6′ × 4′, celestial coordinates (12343n0812), constellation **Virgo**. SBc galaxy; S-shape.
NGC 4552 (M89)	magnitude 11, angular size 2′ × 2′, celestial coordinates (12357n1233), constellation **Virgo**. E galaxy.
NGC 4559	magnitude10.5, angular size 10′ × 3′, celestial coordinates (12360n2758), constellation **Coma Berenices**. Sc galaxy.
NGC 4569 (M90)	magnitude 11.1, angular size 7′ × 2.5′, celestial coordinates (12368n1310), constellation **Virgo**. Sb galaxy.
NGC 4579 (M58)	magnitude 10.5, angular size 4′ × 3.5′, celestial coordinates (12377n1149), constellation **Virgo**. Sb galaxy.
NGC 4621 (M59)	magnitude 11, angular size 2′ × 1.5′, celestial coordinates (12420n1139), constellation **Virgo**. E3/E4 galaxy.
NGC 4649 (M60)	magnitude 10, angular size 3′ × 2.5′, celestial coordinates (12437n1133), constellation **Virgo**. E1/E2 galaxy.
NGC 4631	magnitude 9.7, angular size 12.5′ × 1.2′, celestial coordinates (12421n3232), constellation **Canes Venatici**. Sc galaxy; edge-on.
NGC 4656/57 (M9)	magnitude 11, angular size 19.5′ × 2′, celestial coordinates (12440n3210), constellation **Canes Venatici**. Irr. galaxy; interacting NGC 4657 creates curved tip.
NGC 4699	magnitude10.3, angular size 3′ × 2′, constellation **Virgo**. Sa/Sb galaxy.
NGC 4725	magnitude10.5, angular size 7.5′ × 4.8′, celestial coordinates (12504n2530), constellation **Coma Berenices**. SBb galaxy.
NGC 4736 (M94)	magnitude 8.9, angular size 5′ × 3.5′, celestial coordinates (12509n4107), constellation **Canes Venatici**. Sb galaxy; very bright, no detail.

NGC 4753   magnitude10.6, angular size 2.8′ × 2′,
                celestial coordinates (12524s0112), con-
                stellation **Virgo**. Irr./E galaxy.

NGC 5005   magnitude10.8, angular size 4.1′ × 1.6′,
                celestial coordinates (13109n3703),
                constellation **Canes Venatici**. Sb galaxy;
                bright.

NGC 5033   magnitude10.3, angular size 8′ × 4′,
                celestial coordinates (13134n3636),
                constellation **Canes Venatici**. Sb galaxy.

NGC 5055   magnitude 9.8, angular size 9′ × 4′,
(M94)   celestial coordinates (13158n4202),
                constellation **Canes Venatici**. Sb galaxy;
                bright.

NGC 5102   magnitude10.8, angular size 6′ × 2.5′,
                celestial coordinates (13220s3638), con-
                stellation **Centaurus**. SO galaxy.

NGC 5194/95   magnitude 8.7, angular size 10′ × 5.5′,
(M51)   celestial coordinates (13299n4712), con-
                stellation **Canes Venatici. Whirlpool
                Galaxy**; Sc galaxy. I would have liked to
                have included this legendary sight in my
                primary listing, but it is a difficult object
                in the suburbs. On a really good night,
                with a light filter, I can just detect the
                spiral structure; amazingly, for a face-
                on spiral, traces can also be made out
                in the video image in Figure 8.15.

Figure 8.15.
NGC 5194/95 (M51):
video frame: intensifier.

Figure 8.16.
NGC 5907: video
frame: intensifier.

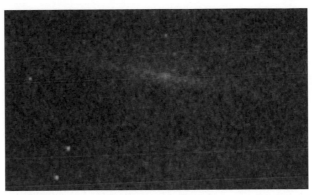

NGC 5253	magnitude 10.8, angular size 4′ × 1.5′, celestial coordinates (13399s3139), constellation **Centaurus**. E galaxy.
NGC 5457 (M101)	magnitude 9, angular size 22′ × 20′, celestial coordinates (14032n5421), constellation **Ursa Major**. Sc galaxy; face-on; spiral structure faintly detectable on good nights with low powers and unintensified viewing.
NGC 5907	magnitude 11, angular size 11′ × 0.6′, celestial coordinates (15159n5619), constellation **Draco**. **Splinter Galaxy**; Sb galaxy; edge-on, needle-shape with some mottling and dust obscuration possible with larger apertures (Figure 8.16).
NGC 6503	magnitude 11, angular size 4.8′ × 1′, celestial coordinates (17494n7009), constellation **Draco**. Sb galaxy.
NGC 6822	magnitude 11.2, angular size 20′ × 10′, celestial coordinates (19449s1448), constellation **Sagittarius**. **Barnard's Galaxy**; Irr. dwarf galaxy like the Small Magellanic Cloud in essence, but appears more like a small, sparse cluster; the planetary nebula NGC 6818 is in the same field.
NGC 7331	magnitude 10.4, angular size 10′ × 4′, celestial coordinates (22371n3425), constellation **Pegasus**. Sb galaxy; thick dust belt on one side (Figure 8.17). It has a small companion galaxy:

Figure 8.17.
NGC 7331: video frame: intensifier. Small companion galaxy NGC 7335.

NGC 7335   angular size 45″ × 25″, constellation **Pegasus**.

NGC 7793   magnitude 9.7, angular size 6′ × 4′, celestial coordinates (23578s3235), constellation **Sculptor**. Sd galaxy.

NGC 7814   magnitude 12, angular size 5′ × 1′, celestial coordinates (00033n1609), constellation **Pegasus**. Sa/Sb galaxy; fine and prominent equatorial dust lane, but a very difficult suburban object.

Planetary Nebulae

The following represent additional planetaries, outside the listing in Chapter 7, and most likely to provide successful viewing from the suburbs. Those with magnitudes less than 12 are probably not bright enough to show well in any respect, and those too spread out and diffuse usually have too low a surface brightness to make effective viewing. Generally as before, I have not specified which observing method is preferable (with an image intensifier, narrowband filter, or no accessory at all) since so much depends on the conditions of your own neighborhood. Experimentation is your best guide, but it is fair to say I have found that the fainter and larger the nebula, the less likely it is to respond favorably to image intensification, and up to a certain point the more it is likely to respond well to a narrowband filter.

NGC 246	magnitude 8.5, angular size 4′ × 3.5′, celestial coordinates (00470s1153), constellation **Cetus**. Large and diffuse.
IC 289	magnitude 12 (15 mag. central star), angular size 45″ × 30″, celestial coordinates (03103n6119), constellation **Cassiopeia**.
NGC 1360	magnitude 9 central star, angular size 6′ × 4′, celestial coordinates (03333s2551), constellation **Fornax**. Very diffuse.
IC 351	magnitude 11 (15 mag. central star), angular size 8″ × 6″, celestial coordinates (03475n3503), constellation **Perseus**.
NGC 1501	magnitude 12 (13.5 mag. central star visible with INT), angular size 55″ × 48″, celestial coordinates (04070n6055), constellation **Camelopardalis**.
NGC 1514	magnitude 11 (10 mag. central star), angular size 120″, celestial coordinates (04092n3047), constellation **Taurus**.
NGC 1535	magnitude 9 (11.5 mag. central star), angular size 20″ × 17″, celestial coordinates (04142s1244), constellation **Eridanus**.
NGC 2022	magnitude 12 (14 mag. central star), angular size 25″, celestial coordinates (05421n0905), constellation **Orion**.
IC 2149	magnitude 10 (14 mag. central star), angular size 10″, celestial coordinates (05563n4607), constellation **Auriga**.
NGC 2371/2	magnitude 12.5, angular size 50″ × 30″, celestial coordinates (07256n2929), constellation **Gemini**. A faint but interesting planetary, which because of two bright zones gives the impression of double ends, and hence the double designation (Figure 8.18).
NGC 2438 (M46)	magnitude 10 (17 mag. central star), angular size 65″, celestial coordinates (07418s1444), constellation **Puppis**.
IC 3568	magnitude 11.6, angular size 18″, celestial coordinates (12329n8233), constellation **Camelopardalis**.
NGC 3587 (M97)	magnitude 11, angular size 150″, celestial coordinates (11148n5501), constel-

Figure 8.18.
NGC 2371/2: video
frame: intensifier
(faintly discernible at
the center of the frame).

lation **Ursa Major. Owl Nebula**; visible with NBF, but surface brightness is very low, making the "Owl" likely to be disappointing in the suburbs.

NGC 4361 magnitude 10.5 (13 mag. central star), angular size 80″, celestial coordinates (12245s1848), constellation **Corvus**.

IC 4406 magnitude 11, angular size 100″ × 35″, celestial coordinates (14224s4409), constellation **Lupus**

NGC 6026 magnitude 12.5, angular size 50″, celestial coordinates (16014s3432), constellation **Lupus**. Ring nebula.

NGC 6058 magnitude 12, angular size 25″ × 20″, celestial coordinates (16044n4041), constellation **Hercules**.

IC 4593 magnitude 11, angular size 13″ × 10″, celestial coordinates (16122n1204), constellation **Hercules**.

NGC 6153 magnitude 11.5, angular size 20″, celestial coordinates (16315s4015), constellation **Scorpius**.

NGC 6210 magnitude 9.7 (12.5 mag. central star), angular size 20″ × 16″, celestial coordinates (16445n2349), constellation **Hercules**. Some detail; oval shape.

IC 4634 magnitude 12 (17 mag. central star), angular size 20″ × 10″, celestial coordinates (17016s2150), constellation **Ophiuchus**.

NGC 6309   magnitude 11.5, angular size $20'' \times 10''$, celestial coordinates (17141s1255), constellation **Ophiuchus**. Upper segment and 14 mag. central star visible with INT.

NGC 6337   magnitude 13, angular size $38'' \times 28''$, celestial coordinates (17223s3829), constellation **Scorpius**.

NGC 6567   magnitude 11.5 (15 mag. central star), angular size $11'' \times 7''$, celestial coordinates (18137s1905), constellation **Sagittarius**.

NGC 6572   magnitude 9.5 (12 mag. central star), angular size $15'' \times 12''$, celestial coordinates (18121n0651), constellation **Ophiuchus**. Twisted appearance of main core.

NGC 6629   magnitude10.5 (13.5 mag. central star), angular size $15''$, celestial coordinates (18257s2312), constellation **Sagittarius**. Pale disk.

NGC 6751   magnitude12 (13 mag. central star), angular size $20''$, celestial coordinates (19059s0600), constellation **Aquila**. Faint but visible with INT; clearly defined oval shape.

NGC 6781   magnitude 12.5 (15.5 mag. central star), angular size $105''$, celestial coordinates (19184n0633), constellation **Aquila**.

NGC 6857   angular size $40''$, celestial coordinates (20019n3331), constellation **Cygnus**. Nebula invisible with INT.

NGC 6818   magnitude10 (15 mag. central star difficult), angular size $22'' \times 15''$, celestial coordinates (19440s1409), constellation **Sagittarius**. Mottled disk framed by triangle of stars (Figure 8.19); near Galaxy NGC 6822: 11.2 mag.

NGC 6886   magnitude11 (16.5 mag. central star), angular size $9'' \times 6''$, celestial coordinates (20127n1959), constellation **Sagitta**.

NGC 6891   magnitude10 (11 mag. central star), angular size $15'' \times 7''$, celestial coordinates (20152n1242), constellation **Delphinus**.

Figure 8.19.
NGC 6818: video
frame: intensifier.

NGC 6905	magnitude12, angular size 44″ × 38″, celestial coordinates (20224n2005), constellation **Delphinus**. **Blue Flash Nebula**; disk with 14 mag. central star; partially framed by four prominent stars (Figure 8.20).
NGC 7026	magnitude12 (15 mag. central star; right star adjacent), angular size 25″ × 16″, celestial coordinates (21063n4751), constellation **Cygnus**. Appears visually like an elongated smudged double spot (Figure 8.21).

Figure 8.20. NGC 6905: video frame: intensifier. Look carefully near the center.

Figure 8.21. NGC 7026: video frame: intensifier.

NGC 7048	magnitude 11, angular size 60″ × 50″, celestial coordinates (21142n4616), constellation **Cygnus**. Relatively difficult object with 18 mag. central star.
NGC 7293	magnitude 6.5, angular size 12′, celestial coordinates (22296s2048), constellation **Aquarius**. **Helix Nebula**. Unfortunately, the famous Helix Nebula is overall just too spread out and faint to be included on my list of best objects; it is certainly possible to see it, but spectacularly? No.
NGC 7354	magnitude 13 (16.5 mag. central star visible in larger apertures with INT), angular size 30″, celestial coordinates (22404n6117), constellation **Cepheus**.
IC 1470	magnitude 12, angular size 70″ × 45″, celestial coordinates (23052n6015), constellation **Cepheus**. Fan-like; irregular shape.

Emission, Reflection and Dark Nebulae

Reflection nebulae do not usually respond favorably to image intensifiers, because the light that leaves them is usually at the blue end of the spectrum. To have any real chance of seeing them effectively in city conditions, you will need the use of narrowband light pollution filters. Success with emission nebulae will depend on the specific wavelengths of light they are generating, and they often respond very favorably to image intensification. Some nebulae may respond to both forms of viewing (narrowband filters and intensification) and in different ways. Your best chances of discerning any of these sights are on nights of maximum transparency. An infrared band-pass filter for your intensifier may enhance some of the emission nebulae, from areas of low humidity. You might also try locating some of the larger, more diffuse nebulae with the non-magnifying primary lens attachment for the Collins I_3, or equivalent, and this will make many of these objects stand out against the night background in city confines where your telescope fails. Unstable air is not of great significance as generally we will be using low magnifications.

The so-called dark nebulae are difficult objects for us since they will always be affected by light pollution; they need truly dark, transparent skies. Nevertheless, with persistence on really good nights some of these dark nebulae are indeed visible. Where they are surrounded by star fields, image intensification will bring out the contrasting darkness to considerable effect.

A word of caution: no matter what equipment you are using, many nebulae may prove visually disappointing in light polluted areas, but all are worth the effort of trying to observe. There may be some real surprises amongst them. Most likely your best chances are with non-intensified viewing, and light filters that will likely produce at least some success include Orion's Ultrablock or other narrowband filters (Lumicon's O III, Hydrogen beta filters, or similar, work slightly better on only a limited range of nebulae).

NGC 281	angular size 23′ × 27′, celestial coordinates (00528n5636), constellation **Cassiopeia**. **Pac Man** emission nebula and cluster.
NGC 1333	angular size 9′ × 5′, celestial coordinates (03293n3125), constellation **Perseus**. Reflection nebula.
NGC 1499	magnitude 4.5, angular size 145′ × 40′, celestial coordinates (04007n3637), constellation **Perseus**. **California Nebula**; emission nebula, extensive, primarily photographic or visible with large binoculars; difficult to see with telescopes in city confines; possible with a non-magnifying objective in addition to an intensifier.
NGC 1788	angular size 5′ × 3′, celestial coordinates (05069s0321), constellation **Orion**. Reflection nebula.
NGC 1977	angular size 40′ × 45′, celestial coordinates (05351s0444), constellation **Orion**. **Running Man Nebula**; bright reflection nebula adjacent to M42.
NGC 1931	angular size 4′ × 4′, celestial coordinates (05314n3415), constellation **Auriga**. Emission and reflection nebula; compact

nebulosity reminiscent of the Trapezium, with four illuminating stars.

NGC 1999	angular size 16' × 12', celestial coordinates (05365s0642), constellation **Orion**. Emission and reflection nebula with 10 mag. star.
IC 443	angular size 50' × 40', celestial coordinates (05410s0224), constellation **Orion**. Emission nebula; includes the famous **Horsehead Nebula**, B 33; very difficult or impossible from the city. With:
NGC 2024	angular size 20', celestial coordinates (05407s0227), constellation **Orion**. **Flame Nebula**. Emission nebula connected to the Horsehead region; similarly problematic from city locations.
NGC 2068 (M78)	magnitude 8, angular size 8' × 6', celestial coordinates (05467n0003), constellation **Orion**. Bright reflection nebula.
NGC 2237	angular size 80' × 60', celestial coordinates (06323n0503), constellation **Monoceros**. **Rosette Nebula**; reflection nebula, with cluster NGC 2440; see Open Clusters. Central hole visible with narrowband filter – ideal with lowest power possible.
NGC 6888	angular size 18' × 12', celestial coordinates (20120n3821), constellation **Cygnus**. **Crescent Nebula**; emission nebula, faintly visible.
NGC 6726/9	angular size 9' × 7', celestial coordinates (19017s3653/19019s3657), constellation **Corona Australis**. Small, bright reflection and emission nebulae.
NGC 7000	angular size 100', celestial coordinates (20588n4420), constellation **Cygnus**. **North American Nebula**; vast emission nebula; difficult to see with telescopes in city confines; quite possible though, with non-magnifying objective in addition to intensifier.
NGC 7023	angular size 18', celestial coordinates (21005n6810), constellation **Cepheus**. One of the brightest reflection nebulae, with dark lanes.

NGC 7635	angular size 205″ × 180″, celestial coordinates (23207n6112), constellation **Cassiopeia. Bubble Nebula**; emission nebula – possibly a planetary; with 8 mag. central star. Hard to see in suburbs.
S 2-240	angular size 2° × 3°, celestial coordinates (05360n2800), constellation **Taurus.** Supernova remnant – traces detectable from suburbs with sufficient aperture.
IC 405	angular size 18′ × 30′, celestial coordinates (05162n3416), constellation **Auriga.** Reflection and emission nebula, with variable star AE Aurigae.
IC 410	angular size 20′, celestial coordinates (05226n3331), constellation **Auriga.** Emission nebula surrounding cluster NGC 1893.
IC 443	angular size 25′ × 5′, celestial coordinates (06169n2247), constellation **Gemini.** Emission nebula, curved arc; supernova remnant.
IC 5067	angular size 80′, celestial coordinates (20469n4411), constellation **Cygnus. Pelican Nebula**; emission nebula, primarily photographic; possible to see using the same approach as with nearby NGC 7000, though fainter.
IC 5146	angular size 12′ × 10′, celestial coordinates (21534n4716), constellation **Cygnus. Cocoon Nebula**; emission nebula, low brightness; difficult; primarily photographic.
B 143	angular size 30′, celestial coordinates (19414n1101), constellation **Aquila.** Celebrated irregular dark nebula in middle of star field.
B 72	angular size 30′ celestial coordinates (17235s2338), constellation **Ophiuchus.** Famous S-shape dark nebula, more difficult visually than B 143. Use lowest power.
B 86	angular size 4.5′ × 3′, celestial coordinates (18030s2753), constellation **Sagittarius.** Striking dark nebula near

the edge of NGC 6520; quite easy to observe; see also Open Clusters section.

B 92   angular size 15′ × 10′, celestial coordinates (18155s1814), constellation **Sagittarius**. Prominent dark nebula near the edge of the Small Sagittarius Star Cloud.

Chapter 9

Supplementary Catalog for the Southern Hemisphere

No survey of the heavens would be complete without including complete reference to the remaining outstanding celestial sights of the Southern Hemisphere, which are comparable in viewing potential to those featured already in the book. Since a large proportion of Southern sights are visible from my location at latitude 34º north, they are already included and detailed in Chapters 7 and 8. However, there are some wonderful Southern sights of which the author is unable to claim firsthand knowledge, not having been observed through his 18-inch telescope and the equipment discussed in this book. Some of these are amongst the most spectacular known. This supplementary catalog of additional Southern objects is presented to indicate those likely to provide good results from the suburbs; they are necessarily included without illustrations and with limited viewing comments. In terms of potential for real time viewing, they also have not been separated in the manner of the last two chapters. Nevertheless, my intention is to provide full and comprehensive coverage for all residents of planet Earth, and it is hoped that observers in other latitudes will forgive the briefer nature of these objects' inclusion.

Open Clusters

NGC 2547   magnitude 5.5, angular size 15′, celestial coordinates (0810s491), constellation **Vela**.

NGC 3532   angular size 60′, celestial coordinates (1106s584), constellation **Carina**. Fine sight at low powers; contains stars of 8–12 mag.

NGC 4755   angular size 10′, celestial coordinates (1253s603), constellation **Crux**. **Jewel Box**; adjacent to the **Coal Sack** dark nebula; closely placed bright stars of 6–10 mag.

Globular Clusters

NGC 104   magnitude 4.5, angular size 25′, celestial coordinates (0024s720), constellation **Tucana**. **47 Tucanae**, second only to Omega Centauri.

NGC 362   magnitude 6, angular size 10′, celestial coordinates (0103s705), constellation **Tucana**.

NGC 2808   magnitude 6, angular size 7′, celestial coordinates (0912s645), constellation **Carina**. Fine object.

NGC 4372   magnitude 8, angular size 18′, celestial coordinates (1226s724), constellation **Musca**. Mostly 12 mag. stars.

NGC 5986   magnitude 8, angular size 5′, celestial coordinates (1546s374), constellation **Lupus**.

NGC 6352   magnitude 9, angular size 8′, celestial coordinates (1725s482), constellation **Ara**. Many fine stellar points.

NGC 6362   magnitude 8, angular size 9′, celestial coordinates (1732s670), constellation **Ara**. Stellar population similar to NGC 6362.

NGC 6397   magnitude 7, angular size 19′, celestial coordinates (1740s534), constellation **Ara**. One of the nearest globulars; majority of stars 10 mag.

NGC 6541   magnitude 6, angular size 6′, celestial coordinates (1808s434), constellation **Corona Australis**.

NGC 6752   magnitude 7, angular size 15′, celestial coordinates (1911s595), constellation **Pavo**. Outstanding cluster.

Galaxies

NGC 292   magnitude 1.5, angular size 3.5°, celestial coordinates (0053s725), constellation **Tucana. Small Magellanic Cloud**; Irr. galaxy.

NGC 1553   magnitude 10.2, angular size 3′ × 2′, celestial coordinates (0416s554), constellation **Dorado**. SO galaxy.

NGC 1566   magnitude 10.5, angular size 5′ × 4′, celestial coordinates (0420s545), constellation **Dorado**. Sb galaxy.

Large Magellanic Cloud (no NGC designation)
magnitude 1, angular size 6°, celestial coordinates (0520s690), constellation **Dorado**. Irr. galaxy; contains many objects: see Burnham.

NGC 4945   magnitude 9.2, angular size 15′ × 2.5′, celestial coordinates (1305s492), constellation **Centaurus**. SBc galaxy; edge-on; outstanding.

NGC 5102   magnitude 10.8, angular size 6′ × 2.5′, celestial coordinates (1322s363), constellation **Centaurus**. SO galaxy.

NGC 6744   magnitude 10.6, angular size 9′ × 9′, celestial coordinates (1909s635), constellation **Pavo**. SBc galaxy.

Planetary Nebulae

NGC 5307   magnitude12, angular size 15″ × 10″, celestial coordinates (1351s511), constellation **Centaurus**. Use higher powers.

NGC 6326   magnitude12, angular size 15″ × 10″, celestial coordinates (1720s514), constellation **Ara**. Use higher powers.

Emission, Reflection and Dark Nebulae

NGC 2070   angular size 20′, celestial coordinates (0538s690), constellation **Dorado. Tarantula Nebula**; extraordinary object within the Large Magellanic Cloud; rivals the great M42.

NGC 3372   angular size 80′ × 85′, celestial coordinates (1044s595), constellation **Carina. Keyhole Nebula**; magnificent sight, with numerous dark lanes crossing, much in the manner of the Trifid Nebula; contains the famous variable star Eta Carinae.

Coal Sack   dark nebula adjacent to the Jewel Box Cluster, NGC 4755.

NGC 5189   angular size 185′ × 130′, celestial coordinates (1333s655), constellation **Musca**.

Chapter 10

Postscript

You may want to consider a few additional thoughts before making some of your final decisions on the equipment most suited to your needs. Assuming most of your routine observing will be at least from a somewhat city-bound location, it would be easy to pay little attention to the potential for your telescope to be moved readily. Earlier in this book, I made reference to my own desire for such portability, and although you may feel that this will not be so important to you, I am certain that times will present themselves when you wish to view the skies from better sites. Contemplate the potential for viewing what you already have accessed from the suburbs all over again with new-found awe (imagine what an image intensifier and narrowband light filters will do in these more ideal surroundings!), but also all the new objects that will become possible to view. So it is wise to carefully reflect on the ability you will have to move easily whatever equipment you select, a task that you should make sure is not so strenuous and awkward that you are not inspired to go through it too often.

From your suburban site, it is unlikely that you will have an unrestricted horizon, which again speaks to the need for some degree of portability, even at your home, for your telescope. Then the ever-present problem of intruding lights in one direction or another also may necessitate moving the scope to different places to accommodate these inconveniences. I always seem to be dealing with some light source or another, and have found many ways to dodge them. (This sometimes even involves hanging sheets on extending lines.)

But it helps simply to be able to move the telescope easily. It goes almost without saying that a permanent telescope housing probably is not realistic for many people in our situation. I must admit that having such a facility of my own was always a dream. If you have read or looked through many journals and literature on amateur astronomy, you will have already noticed the wonderful observatories that have been built or purchased for such private use. What amateur astronomer wouldn't want such a place for him or herself? However, I have come to realize that it would only hamper what I am able to do from my home location, and therefore have accepted the reality of my circumstances. Nevertheless, what I have settled for has indeed turned out to be a good alternative. Before you make any decision on a permanent housing, take the time to consider your complete situation, and how you might actually be hampered by having a such a facility. You will be surprised how a small area of sky, the very one that you may have concluded you could live without, turns out to be the region that is critical for certain objects and viewing times.

Having weighed all of this, and if you have come to the same conclusion as me, it is extremely beneficial to set up the mounting, equatorially aligned, in the various locations the telescope is to be used, and mark or construct exact positioning guides. This way, it will be possible for quick placement at a moment's notice. Only at those times when absolutely perfect alignment is necessary (i.e. for astrophotography), will you have to do more than a minute's work. The difference between these quick-alignment placements and exact polar alignment will be found to be very slight. It will repay your set-up efforts many times, every time you use the telescope, to be very close to true alignment.

Much has been written about the need for dark adaptation for our eyes, in order to be in a position to see the maximum that is possible. William Herschel was known to go into complete darkness for at least twenty minutes prior to any viewing session, and nobody can argue that he did not extract all that he could from his vision. In our local suburban environment, we may still wish for this ideal, but we have to realize again that such ultimate dark adaptation will not be possible. Add to this the problem of moving automobile lights, overhead planes and helicopters, bright reflected light from malls, sports arenas and commercial centers, lights from neighboring houses and their backyard security lights; the list goes on. Any one of these factors is enough to ruin any true dark adaptation, although some degree of it is still possible with care. Your best bet will typically be in the small hours of the morning, after most people are asleep, businesses are closed, cars are garaged, planes have landed, and the world is at its most tranquil. You will now stand the best chance of experiencing the maximum that deep space has to offer, although the use of an image intensifier will prevent deep attainment of dark adaptation. It is only when you compare the sheer visual punch of these devices against normal viewing that you will realize exactly what you are dealing with. The field of view is literally glowing with light. All of this is not necessarily a disadvantage, however, since at our home location, at least, the need for significant dark adaptation is hardly necessary while we are using an intensifier. Since we will not achieve it anyway, the issue is moot. It is only when we switch to regular or filtered viewing that we may wish

Figure 10.1. The author, at home, prepares for a night of viewing with his 18-inch JMI Newtonian Reflector.

for truer dark adaptation. It is also a fact that as long as we switch back and forth from one form of viewing to the other, it is unrealistic to even consider attaining it. You should not therefore expect to have nearly the same degree of visual acuity with non-intensified viewing while you are observing in these types of mixed sessions.

Because of the whims of my local climate, it is often impractical for me to attempt much in the way of galactic and certain other deep space observing during late spring, a good portion of the summer and early fall. At the coast in Southern California we are frequently plagued with a night-time marine layer and haze during these times. In order to see all that the sky has to offer, I frequently spend long sessions at the telescope whenever the air is as good as it can be, enough so that the sky can move through a large part of the celestial sphere. Sometimes this results in all-nighters! Your own circumstances presumably differ from mine, and even if you do not intend to spend entire nights occupied with looking at the sky, you will soon come to realize how readily this tends to happen anyway, once you have focused viewing objectives and a sustained clear sky. So why not simply periodically plan it that way, and allow yourself the time necessary to look properly and completely at your selected objects? It is such a waste of any good viewing opportunity not to take advantage of it. If you do not intend a lengthy session, it is better to plan on viewing only a small selection of objects properly rather than dashing madly around the skies.

If, one night, I have planned to spend most of the night hours with my telescope (this often happens after a long spell of night-time marine layer "cabin fever"), I have found a method that gets me through the night without too many ill effects, and even through the next day! This is significant to me since I have always been one to want my sleep. As you complete your viewing of a particular section of the sky, take a catnap on a nearby couch, with your head propped up, as you wait for a new segment of the sky to rise. Keep some low light on (you don't want to fall into deep sleep), and don't get too comfortable (as in sleeping in bed), and you will find that you "wake up" quite refreshed, usually within half an hour to an hour. If you don't trust yourself to do this, a minute timer might do the job. Just repeat this procedure throughout the night; it becomes remarkably easy to "pull an all-nighter", and still be functional the next day. The stimulation of all you are seeing will stoke the fires as well, and will make this nocturnal activity not too difficult. Don't do this two nights in a row, however! Happy viewing!

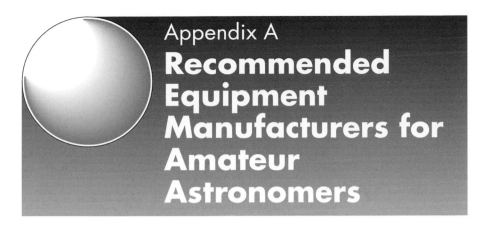

There are numerous manufacturers of outstanding astronomical equipment today, too many to list. The following represent the author's own preferred companies, carefully considered to be sure, but by no means meant to imply that other outstanding manufacturers do not exist. Whenever possible, I have listed some European distributors of American companies. In other instances, I have indicated whether a particular company handles exports if they do not have an overseas distributor.

Telescopes

1. JMI (Superior split-ring equatorial Newtonian truss telescopes; accessories for all makes, including motorized focus units)
   810 Quail St., Unit E
   Lakewood, Colorado 80215
   (303) 233-5353
   www.jimsmobile.com

   In Europe, JMI products are available through:

   Broadhurst, Clarkson & Fuller
   6, Tunbridge Wells Trade Park
   Lonfield Road
   Tunbridge Wells, Kent
   England TN2 3QF

(44) 2074-052156
www.telescopehouse.co.uk

Intercon Spacetec
Gablinger Weg 9
D-86154 Augsburg 1
Germany
(49) 8214-14081
www.intercon-spacetec.com

La Maison de l'Astronomie
Devaux-Chevet
33-35, rue de Rivoli
Paris, France 75004
(31)1427-79955
www.maison-astronomie.com

2. Obsession Telescopes (Superior Dobsonians)
   PO Box 804s
   Lake Mills, Wisconsin 53551
   (920) 648-2328
   www.obsessiontelescopes.com

   Obsession sells and exports direct worldwide, but has no
   distributors.

3. Astro-Physics (Superior apochromatic refractors)
   11250 Forest Hills Rd. Rockford, Illinois 61115
   (815) 282-1513
   www.astro-physics.com

   Astro-Physics exports direct to Europe and many other
   countries, but does not export to countries where there is
   a distributor. Distributors include:

   Baader Planetarium KG
   Thomas Baader
   zur Sternwarte
   82291 Mammendorf
   Germany
   (081) 458802

   Medas SA
   57, Avenue P. Doumer
   BP 2658
   03206 Vichy Cedex
   France
   (04) 70-30-19-30
   www.medas.fr

   Unitron Italia Srl.
   Giovanni Quarra
   via Agostino Lapini, 1
   50136 Firenze
   Italy
   (055) 667065
   http://www.untronitalia.it

4. Parks Optical (Superior traditional Newtonians and other types)
    PO Box 716
    Simi Valley, California 93062
    (805) 522-6722
    www.parksoptical.com

    Parks telescopes are available in England through:

    Venturescope
    The Wren Centre
    Westbourne Road
    Emsworth, on the Hampshire–Sussex border
    England PO10 7RW
    www.telescopes.co.uk/parks.htm

5. Orion (Maksutov-Newtonians, achromatic & apochromatic refractors, and low-cost, high quality Dobsonian telescopes up to 10-inch aperture, eyepieces, filters and accessories)
    PO Box 1815-S
    Santa Cruz, California 95061
    (800) 676-1343
    www.telescope.com

    Orion does not directly export its products internationally, but does have two distributors in England:

    SCS Astro
    The Astronomy Shop
    1 Tone Hill
    Wellington, Somerset
    England TA21 OAU
    (44) 1823-665510
    www.scsastro.co.uk

    Broadhurst, Clarkson & Fuller (see JMI)

6. Astro Works Corporation (sophisticated observatory instruments)
    PO Box 699
    Agulia, Arizona 85320
    (520) 685-5000
    www.astroworks.com

    Astro Works is a small company, that has no e-mail contact. Contact them by phone for possible export arrangements

7. Takahashi (premium apochromatic refractors; reflectors, catadioptrics)
    Texas Nautical Repair, Inc.
    3110 S. Shepherd St.
    Houston, Texas 77098
    (713) 529-3551
    www.LSSTNR.com

Takahashi products are available at dealers throughout the world, and in England from:

True Technology Ltd.
c/o Nick Hudson
Woodpecker Cottage
Red Lane
Aldermaston, Berks
England RG7 4PA
(44) 01189-700-777
www.trutek-uk.com

8. Internet Telescope Exchange (custom Maksutov–Newtonians; apertures to 16 inches)

3555 Singing Pines Road
Darby, Massachusetts 59829
(406) 821-1980
www.burnettweb.com/ite

ITE ships worldwide; its products are also available in England through:

SCS Astro (see Orion)

Typical Approximate Costs in US (US$) for New Equipment

Dobsonian reflectors
6–10 inch: $500–$2,000
12–30 inch: $1000–$10,000

Equatorial Newtonians
6–10 inch: $500–$3000
12–18 inch: $1250–$13,000

Other types of equatorial reflector (i.e. classical Cassegrain)
18–30 inch: $20,000–$75,000 + up

Equatorial achromatic refractors
3–6 inch: $400–$3000

Equatorial apochromatic refractors
3–5 inch: $1500–$5000
6–7 inch: $5000–$12,000
8–10 inch: $20,000–$40,000

Equatorial Maksutov–Newtonian
6 inch: $2500
8 inch: $5000
10 inch: $10,000

Schmidt–Cassegrain
8–12 inch: $1000–$4000
16 inch: $15,000

Light Pollution Filters, Eyepieces and Accessories

1. Lumicon
   6242 Preston Avenue
   Livermore, CA 94550
   (925) 447-9570
   www.lumicon.com

   Lumicon exports its products throughout Europe, but also has dealers in the UK and most European countries. In Europe, contact:

   Broadhurst, Clarkson & Fuller (see Telescopes: JMI)

2. Orion (see Telescopes)

3. Televue Optics (Eyepieces, superlative small refractors, other accessories)
   100 Route 59
   Suffern, NY 10901
   (845) 357-9522
   www.televue.com

   Televue products are available widely in Europe. Contact:

   SCS Astro (see Telescopes: Orion)
   Broadhurst, Clarkson & Fuller (see Telescopes: JMI)
   Venturescopes (see Telescopes: Parks)

4. Collins Electro Optics (Specialized light pollution filters for intensifiers, video cameras, frame averagers etc. See image intensifiers)

Typical Approximate Costs in US (US$) for New Equipment

Eyepieces:
$25–$400 (depending on type: traditional versus modern multi-element, highly corrected types)

Light pollution filters:
$70–$250 (depending on size: 1_-inch or 2-inch, and type)

Digital setting circles (including encoders):
$350–$500 (depending on presets/functions)

Image Intensifiers

Specialized, ready for astronomical use:

1. Collins Electro Optics (Complete systems for astronomy – no adaptation necessary)
   9025 East Kenyon Avenue
   Denver, CO 80237
   (303) 889-5910
   www.ceoptics.com

   Unfortunately, Collins has no distributors overseas. Presently, certain export restrictions apply; contact Collins Electro Optics for export information for your situation.

Other Image Intensifier Products

The following products will need adaptation.

2. Electrophysics Corporation (Generation II, III, IV units)
   373 Route 46 West
   Fairfield, New Jersey 07004-2442
   (973) 882-0211
   www.electrophysics.com

   Electrophysics exports to many countries; some European distributors include:

   AM Vision
   The Old Schoolhouse
   Wilberfoss, York
   England YO41 5NA
   (0044) 0 1759 388235

   Jabsco
   Ostsrabe 2B
   D022844 Norderstedt
   Germany
   (040) 53533730

   Jenoptec
   12, rue J-B Huet
   Les Metz
   78350 Jouy en Josas
   France
   (33) 01 34659102

3. D & VP Corporation
   PO Box 54074 N. Salt Lake
   Utah 84054-0274
   (801) 299-8548
   www.dandvp.com
   or: www.nightvisionweb.com

D & VP has no European distributors, but exports most of their products.

4. Stano Components, inc.
   PO Box STANO
   Silver City, Nevada 89428
   (775) 246-5281/5283
   www.stano.night-vision.com

   Stano does not export to countries outside USA.

5. Aspect Technology and Equipment, inc.
   811 East Plano Parkway, Suite 110
   Plano, Texas 75074
   (800) 749-3802 / (972) 423- 6008/7717
   www.aspecttechnology.com

   Contact Aspect directly for possible exports and dealers in Europe.

A Sampling of European Companies Supplying High Quality Image Intensifiers

6. Optex (Gen III-equivalent miniature systems, probably ideal for astronomy)
   20–26 Victoria Road
   East Barnet
   Hertfordshire
   England EN4 9PF
   contact: simon@optexint.com

7. The House Of Optics (Russian intensifier units including Gen III)
   Hunstanton
   Norfolk
   England
   07879-214651
   www.houseofoptics.ltd.uk

8. Delft Instruments NV
   Röntgenweg 1, 2624 BD
   PO Box 103, 2600 AC Delft
   The Netherlands
   +31-15-2-601-200
   www.delftinstruments.com

9. Eureca
   Messtechnik Gmbh
   Am Feldgarten 3
   50769 Köln
   Germany
   www.eureca.de

Typical Approximate Costs in US (US$) for New Equipment

Fully dedicated Generation III systems, ready to use (i.e. Collins):
$2000

Generation 2 (intensifier only – will require eye lens and adapters):
$500–$850: more advanced Generation 2 products up to $2000.

Video Cameras For Astronomy and Accessories

1. Adirondack Video Astronomy (also sells image intensifiers)
    26 Graves St.
    Glen Falls, NY 12801
    (518) 812-0025
    www.astrovid.com

    Adirondack's distributor in the UK is:

    True Technology Ltd. (see Telescopes: Takahashi)

2. Santa Barbara Instrument Group
    147-A Castilian Drive
    Santa Barbara, CA 93117
    (805) 571-7244
    www.sbig.com

3. Internet Telescope Exchange (see Telescopes)

Typical Approximate Costs in US (US$) for New Intensifier Accessories

Camera and attachments for intensifier:
$850

Media converters and software for interfacing with computers:
$400

Infrared band-pass filters (for light-polluted areas):
$250

Recursive frame averager:
under $1,000

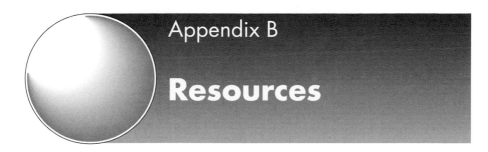

Resources

Prominent Amateur Astronomical Associations and Links

ALPO – Association of Lunar and Planetary Observers
www.lpl.arizona.edu/alpo.com

British Astronomical Association
www.ast.cam.ac.uk/~baa.com

Astronomical Society of the Pacific
www.astrosociety.org
(Website supplies worldwide listing of astronomical organizations, etc.)

International Supernovae Network
www.supernovae.net

American Association of Variable Star Observers
www.aavso.org

International Dark Sky Association
www.darksky.org

Astronomical League
www.astroleague.org

International Meteor Organization
www.imo.net

International Occultations Timing Association
www.lunar-occultations.com

NASA Photographic/Information Reference Website
www. images.jsc.nasa.gov
(A comprehensive catalog of NASA's photographic records of
lunar, planetary and deep space subjects, together with links
to many other relevant sites, including amateur images,
groups, etc.)

Figure B.1. The planet and satellites (and shadow), with Great Red Spot 11/12/2001,
recorded in real time by Astrovid 2000 camera in conjunction with 18″ JMI reflector and
Televue 2× Barlow lens

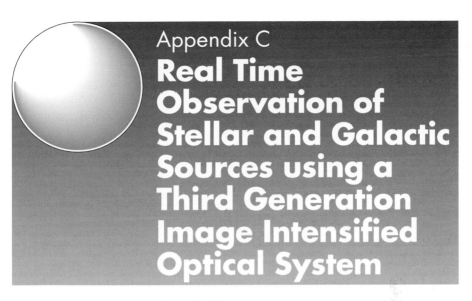

Appendix C

Real Time Observation of Stellar and Galactic Sources using a Third Generation Image Intensified Optical System

W.J. Collins

August 4, 1998
(Reprinted with his kind permission)

Introduction

This report will familiarize the reader with the optical frequency spectrum of astronomical objects for observation using the I_3 intensified optical system, and the system's performance as it relates to the objects' spectrum.

Galaxy Spectra

It is generally agreed that the spectral range of human vision is between ~ 380 and ~760 nanometers (nm = billionths of a meter). Referring to Figure C.1, the A curve represents the spectral response of a typical Generation 3 intensifier (as used in the I_3 Piece). One can immediately see that the tube response extends to 900 nm, with the peak at ~ 775 nm. This region of the spectrum between 760 and 900 nm is included

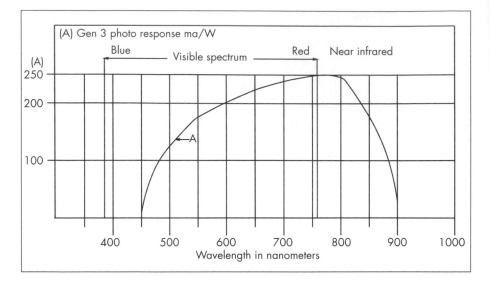

(A) Gen 3 photo response ma/W

Figure C.1. Collins Electro Optics spectral response curves of Generation 3 intensifier.

in the near infrared portion of the electromagnetic spectrum and is not visible to the eye in real time without the assistance of a device such as an image intensifier. Now referring to Figure C.2, curve B, this is a curve-fitted plot of galaxy types (spiral, elliptical, irregular). The slope of curve B is a good fit to the tube response of curve A, particularly between 550 and 800 nm. Also note that the majority of spectral output falls above 700 nm and extends to 900 nm with good uniformity. (Galactic spectra extend well beyond 900 nm; however, this report concerns itself with the spectrum of tube operation $\leq$ 900nm.) Also note that the spectrum of curve B actually begins below the threshold of sensitivity of curve A. Generation 3 devices are essentially blind to this (violet) portion of the visible spectrum. Fortunately, this narrow band between 400 and 450 nm represents a small percentage of the entire galactic spectrum that is visible.

Ellipticals

Curve B represents the average spectrum of the entire galactic mass for the three galaxy types. Within individual galactic types, the spectrum (and hence the intensifier response) can be further quantified. Ellipticals, which are classified by Hubble category EO to E7, depending on how round (EO) or elliptical (E7) they appear, are symmetrical in shape. M87 is an example of an elliptical. The stellar population of ellipticals is called Population 1 from the work of Walter Baade at the 100-inch Mount Wilson telescope. Ellipticals are in fact comprised of "old" Population 1 stars. These are "metal rich"

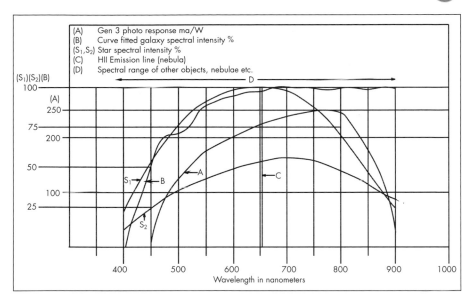

(A)	Gen 3 photo response ma/W
(B)	Curve fitted galaxy spectral intensity %
(S₁,S₂)	Star spectral intensity %
(C)	HII Emission line (nebula)
(D)	Spectral range of other objects, nebulae etc.

Figure C.2. Collins Electro Optics spectral response curves compared with galactic and stellar radiation.

and predominantly M class, including red giants, most of which exceed 10 billion years in age. From a spectral standpoint, ellipticals are very energetic sources in the red and infrared portions. Ellipticals display a uniform spectral curve across their entirety and, when taken individually, their spectra are similar to the S2 curve of Figure C.2 (an M class star). This far red/infrared spectrum makes ellipticals an excellent match to the image intensifier response (curve A).

Spirals

Spiral galaxies can be normal (Hubble type S) or barred (Hubble type SB). Both types are also classified A, B or C, as to the tightness of spiral structure that they display, with A being the tightest and C being most open. Also, the size of the nucleus relative to the spiral structure is A, the largest, and C, the smallest; some galaxies show disk-like structure without spiral structure and are termed SO. Within the nucleus of spiral types, old population stars predominate as in ellipticals. The nuclear bulge is therefore also an excellent match for the intensifier spectral response. As we look into the spiral structure, gas, dust, and young (Population 2) stars are most prominent. This makes the spiral structure more skewed towards the blue portion of the spectrum, making the spiral structure less visible using image intensification than the nuclear central bulge. This can be confirmed observationally by noting the increase in luminosity between the nuclear and the spiral structure. The image intensifier response to the spiral structure, independent of the nuclear bulge, is very dependent on the averaged spectrum that comprises the entire spiral structure. To clarify this important point, spiral galaxies that present their structure to us without oblique perspective, such as M101, will appear highly intensified in the nucleus and will show little difference in visual observation to their spiral structure, due to the predominantly blue response in the spiral arm region. As the observer's plane of view to the galaxy becomes more oblique, the dust lanes become more prominent. Galaxies such as M107 and NGC 4565 present an "edge-on" appearance; the dust lanes have a strong infrared signature, making these types ideal for image intensified observation.

Irregulars

These have Hubble classification IRR1 (mostly O and B type stars and H II regions) with a general lack of dust clouds, and IRR2 (not resolvable into stars, no H II regions) with prominent dust lanes. Of these two, IRR2 types have a more red/infrared spectrum (dust lane infrared signature) and may be a better match to the intensifier spectral sensitivity.

Two additional galaxy types not easily classified are Seyfert (1 and 2) and BL Lacertae objects. Both Seyfert types have unusually small and optically intense (star-like) nuclei. Of the two types, Seyfert objects have a more energetic infrared spectrum. BL Lacertae objects have rapid intensity variations in visible and infrared wavelengths and may be good candidates for image intensified observation.

Stellar Spectra

Referring to Figure C.2, curves S1 and S2, curve S1 is a star with spectral class G, such as our Sun or Capella. Notice the distribution of spectral energy, with the majority in the visible spectrum and decreasing (although still significant) in the infrared. These "main sequence" stars have surface temperatures of ~ 5000 kelvin, producing the spectral distribution curve of S1. Looking at curve S2 in Figure C.1, we see a spectral distribution shifted more towards the red/infrared portion of the spectrum. This falls into spectral class M and includes red supergiant stars such as Betelgeuse in Orion, or Antares. M stars such as these are a fine match to the spectral response of the imaging tube. M stars have surface temperatures in the 3000 kelvin range, causing their red shifted spectrum. Spectral class types B, A and F are not shown. These hotter, bluer stars have spectra shifted towards blue and ultraviolet (the spectral region at the 400 nm end of the chart) and may show modest or no intensity increase when viewed with a Generation 3 intensifier, due to their spectrum falling in the region near the tube's minimum response. K types fall between G and M and are also not shown. Understanding where a star's spectral class falls within the intensifier's effective spectral range (curve A) will allow the user to predict the effectiveness from an image intensification standpoint. M giants and supergiants give the greatest potential for image intensified observability.

Nebulae

Emission Nebulae

Refer to Figure C.2, (D) nebulae along the top section. Notice that the spectral distribution ranges across the entire spectrum shown. The most predominant frequency for the nebulae is centered at the H II line in Figure C.2. This is the H alpha line at 656.32 nm and is the result of spontaneous photon emission from the ionized hydrogen gas present in the nebula as the electrons decay from the third to second energy level. Other gases present within the nebula may also be ionized, as is the case with the Great Nebula in Orion, in which ionized helium and oxygen are also present. These optical recombination lines give rise to other characteristic spectra causing emission lines at other wavelengths. In the case of M42, ionized oxygen (at 500.7 and 495.9 nm) produces the green light present, with H II emission producing the greater part of the red emission.

Emission nebulae will show greatly enhanced observability using Generation 3 intensification when most of their emission spectra occur within the H II region. This is the first emission line in the Balmer series of hydrogen emission lines. As electrons decay from higher valence levels within the hydrogen atom, they emit photons at higher frequencies. This gives rise to the Balmer series of visible emission spectra with the first line (known as hydrogen alpha or H II). There are five emission lines in the Balmer series that are present in the visible spectrum at 656, 486, 434, 410 and 397 nm. As previously stated, the H II line is most observable with Generation 3 intensification, with hydrogen beta (486 nm) also visible. We may therefore predict the image intensified observability of emission nebulae by first knowing what ionized gases constitute their observable spectra and their corresponding emission line frequencies. These emission lines can then be plotted in relation to the tube response curve and their potential for amplification predicted.

Planetary Nebulae

As with emission nebulae, the ionized spectra present are due to their proximity to a star(s), and in the case of planetary types, their surrounding of a hot star (30,000–100,000 kelvin). The Ring Nebula in Lyra is a good example with a characteristic circular shell (hence "planetary", termed by Herschel) surrounding the central star with a temperature of 70,000 kelvin. The strong ionizing radiation gives rise to hydrogen (Balmer series) and oxygen lines at 500.7 and 495.9 nm. The shell of expanding gas in M57 is an excellent choice for Generation 3 intensification because of its H II abundance, and to a lesser extent, its oxygen lines. The central star, although observable with intensification, is nevertheless very blue in color and at the low end of the intensifier response. The observability of planetary nebulae is based on the identical criteria previously stated for emission nebulae.

Reflection Nebulae

Certain nebulae are simply clouds of dust that are illuminated by nearby stars and reflect the stellar spectra present. The Pleiades are a good example of a reflection nebula in the presence of young (hot, blue) stars. The nebulosity present in the cluster reflects the blue spectrum present in these stars. The potential for intensified observability can be determined by the characteristic spectrum of the stars that illuminate reflection nebulae.

Visual versus Silicon-Based Spectral Sensitivity

When used for visual astronomy purposes, image intensified devices are often met with questions such as "why is the image green?". The logical reasoning behind this is as follows:

Color	Red	Orange	Yellow	Green	Blue	Violet
Wavelength, nm	670	605	575	505	470	430
Relative radiant power	10,000	1000	100	1	2	20

Therefore, at 505 nm, the minimum threshold of perceptible vision is 1; yellow light requires 100 times the intensity to produce the equivalent visual response, orange 1000×, red, incredibly, 10,000×, blue 2× and violet 20×. This visual spectral sensitivity is based on scotopic (rod) vision. During photopic (cone) vision (light levels above approximately 10 lux), the peak sensitivity shifts upwards to 555 nm. The image intensifier phosphor screen spectral frequency is centered at 530 nm. With the phosphor screen output illumination level at 2.25 foot lamberts maximum, the visual response falls within the threshold region between scotopic and photopic visual sensitivity. Therefore, 530 nm represents the ideal median frequency for the typical level of visual adaptation that occurs when using a Generation 3 image intensifier. Also, and very importantly, as the intensifier illumination level drops (when imaging a low surface brightness galactic object, for example), the eyes' response becomes predominantly scotopic and the perception of color will actually disappear, because of the retinal rods' insensitivity to color and the visual transition to gray scale. Therefore, the green image present at higher illuminated image levels will become less apparent as the objects' level of illumination decreases, to the point of showing little or no color, and as the tube output approaches the equivalent background illumination (EBI) of the tube. Green frequency phosphor also greatly reduces the power requirement necessary from the tube's power supply because of the much greater visual sensitivity to green 530 nm light, which in turn extends the operating hours with a given (battery) power source.

Tube Spectral Response

Refer now to Figure C.1, Generation 3 photo response. The peak spectral response of the tube is at 775 nanometers. The "gain" of the tube, which is gain = output illumination/input illumination, is independent of photo response. The gain setting for the Generation 3 tube is 50,000. This gain is present across the spectrum of photo response.

This brings us to one of the most important concepts concerning the use of a Generation 3 intensifier for astronomical objects. That is, the ability to dynamically amplify the optical spectrum of a star, nebula, or galaxy is directly related to the integrated spectrum of the object. The same statement applies to human vision except that the peak response is literally at the other end of the spectrum.

An excellent example of the differences between visual and intensifier response is apparent with SC galaxy types, such as M33. The naked eye response does not give the appearance of the galaxy nucleus being brighter than the spiral arms up to the magnitude that is actually measured with instrumentation for bolometric response. This is due to the eye responding with much greater sensitivity to the spiral arm section, made up of bluer stars, than the nucleus, which, although much more energetic than the spiral arms, is nevertheless

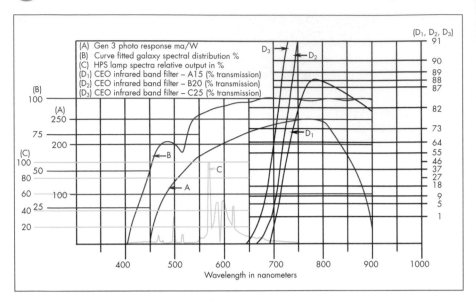

The chart legend reads:

(A) Gen 3 photo response ma/W
(B) Curve fitted galaxy spectral distribution %
(C) HPS lamp spectra relative output in %
(D₁) CEO infrared band filter – A15 (% transmission)
(D₂) CEO infrared band filter – B20 (% transmission)
(D₃) CEO infrared band filter – C25 (% transmission)

Wavelength in nanometers

comprised of much older red M class stars and large H II regions, to which the eyes' response is much less sensitive.

The intensifier responds in a much more linear fashion in comparison to the eye. Let's look at the bandwidth between 505 and 605 nm for the eye and the intensifier. First, for the eye, at 505 nm the response is 1; at 605 nm, it is 1000 (in this case, 1000 is 1/1000 or 0.001 as sensitive as at 505 nm). For the intensifier: at ... 505 nm, it is 135 ma/w; at 605 nm, it is 205 ma/w, a ratio of 205/135 = 1.52:1, versus 1000: 1 for the eye. This illustrates an additional important point: the Generation 3 intensifier provides a more linear and therefore more accurate response to astronomical objects over a large portion of the spectral response of human vision, as compared to the eye over the same spectral range.

Figure C.3. Spectral response curves of Collins Electro Optics band-pass filters.

On the Philosophy of Building versus Buying

As you became aware in Chapter 2, I always had a "thing" regarding the direct connection between the telescope, as an entity in itself, and observing with it. Even so, designing and building the telescope by oneself expands this further. For many years I simply could not separate these two worlds, as there is no doubt in my mind that there is another dimension added to the wonder of it all by having fashioned an instrument oneself. Even more remarkable is the near disbelief one experiences when an assemblage of seemingly unlikely and non-astronomical components actually works when at the inaugural viewing session! There is a profound satisfaction in putting together, in unconventional fashion, a unique, semi-designed apparatus, one which is determined partly out of ingenuity, and partly out of the force of circumstance, as well as luck. Even better is that stunning and memorable occasion when it exceeds one's best hopes. For me, this delight would never go away, even with frequent use of these creations.

Not only was a good deal of the entire observing experience very much connected to this sense of working with this contraption of one's own making, but even in preparing it for use before the session. In the meantime were all the times of planning, with a never-ending stream of improvements and refinements to its design that might be possible to bring about. I cannot pretend that all of this does not have some considerable effect on how one relates to astronomy: it serves to personalize it in a way that cannot be appreciated or experienced any other way. Meanwhile, if you conclude that you (a) have the time needed to pursue home scope-building, (b) have the minimum required facilities to carry it out, or (c) are actually pushed into it by circumstances similar to those that pushed me, you will never have any regrets. However, don't go into it oblivious to the demands that the pursuit will place on you, some brought about through the inexplicable and energizing absorption that will come out of its very being. You might even elect to construct a hybrid of your own design, combining various components you make yourself with some commercial parts.

As a project right out of the box, you would be ill-advised to tackle anything too grandiose or sophisticated. While I believe you should be encouraged to try something on the more adventurous side (I never started small and unassuming, an $8\frac{1}{4}$-inch F8 reflector being my first homebuilt!), design your edifice with some practical considerations in mind. For a scope of its aperture and focal ratio, only the simplest of facilities are needed, and you can expect some good results with very straightforward designs. Maybe the easiest mounting is a yoke-type, made of a wood rectangle with roller bearings at each end of the polar axis, and on both sides for declination. It would be possible to fit this telescope with commercial digital setting circles (probably not an item for home construction – although this has been done, too), as well as a commercial or homebuilt drive on one or both axes.

Remember, there is no such thing as too massive or stable in telescope construction! Designing your own scope presents some special opportunities: in making the above $8\frac{1}{4}$-inch reflector, I used an oversize tube, both in diameter and length. The secondary was therefore well-buried far from the top of the tube, and this limited any stray light or observer-induced air currents from interfering with that part of the optical path. However, I still needed to access the secondary in order to adjust it, something made quite difficult by its very strategic placement. To do this, a door was cut out and hinged adjacent to the mirror, which solved the problem in a very pleasing and efficient way. An easy system for collimation was devised using a novel approach, which I carried over as well onto my largest scope construction (the $12\frac{1}{2}$-inch). From three protruding "ears" connected to the mirror cell I attached (by a rotatable coupling design) three long rods with turning handles at the opposite end. These rods were mounted parallel to the tube through holes in their mounting brackets, which allowed them to turn, and they resembled in some ways the multiple operation rods and turning wheels we see on all the grand old observatory refractors. It was easy to collimate the scope while looking in the eyepiece hole at the same time with no stretching or straining; a very effective solution to an age-old problem. It is surprising to me that this system has not been more widely configured by other builders, and even commercial makers.

The larger sizes present ever-increasing problems. With growing apertures the need to avoid unwieldy tube lengths becomes one of the great mechanical challenges. Simply reducing the focal ratio is not nearly such a ready solution as it may sound, since short ratios and larger primaries also create diametrically increasing optical challenges for the amateur glass worker; the tolerances for accuracy shrink to minute amounts. Both of these factors were significant in my calling it a halt at $12\frac{1}{2}$ inches. In Chapter 2, you may recall that this, my last homebuilt, had a focal ratio of F9, a readily feasible optical surface to fashion at this aperture with the relatively limited means and facilities I had at the time. However, I have to point out that the tube length definitely represented an upper limit for what I was prepared to deal with or risk using in total darkness; it necessitated platforms and rigs that began to approach dangerous extremes. I realized that I had reached my limit, and that any future scopes would have to be made to smaller focal ratios if I wanted to keep the apertures growing.

As a point of interest, I built a large German equatorial mounting for this $12\frac{1}{2}$-inch monster. The main body of it was made out of two massive 6-inch pipe fitting T's, with 6-inch pipe lengths filled with concrete as shafts. The declination axis and polar axis required me to ream out both T's so that the 6-inch pipe could pass through and rotate freely. When greased they provided quite an excellent bearing due to the large surface areas of each. I burned out two electric drills fitted with grinding wheels in achieving that goal, and the finished mounting was indeed highly successful, and a remarkably smooth and responsive affair. Friction or locking on each axis was provided by a large simple right-angled screw, drilled and tapped into each T, and bearing down on the pipe inside by being threaded through the T's. Counterweights were huge disks of lead attached to the 6-inch pipe by locking bands. The tube was all skeletal, and because I had constructed it fully enclosed and baffled at each end, it provided ventilation and light shielding, as well as reasonable protection from operator heat and breathing. Focusing was achieved by moving the focusing unit attached to the secondary on a track along the tube, and not by the eyepiece itself. This was similar to a design in *Telescopes for Skygazing* by Dr. Henry E. Paul. This allowed for a smaller secondary, because the eyepiece can remain closer to the tube. It also featured as well a two-stalk secondary mounting, which appeared as only one because they were in line with the tube when looking from the top. The thin metal that was allowed by this approach gave all the needed sturdiness, as well as a wide latitude of adjustments to the secondary. Also featured was the usual three-point adjustments on the secondary mirror mounting itself, which we are used to encountering. Since I had intended the scope primarily for planetary viewing, the long focal length, together with the secondary's greater proximity to the eyepiece, allowed it to be smaller than the conventional size. The much more limited diffraction, in practical effect zero, unlike that caused by larger secondaries, had to be amongst the reasons it performed so well.

Ultimately, I experimented with a hydraulic drive on the polar axis, but before the time that the telescope was packed away (only to lie dormant in my basement), I did not feel that I had eliminated all the bugs in it, and so can only claim its operation was partially successful. A simple arrangement based on such drives outlined in *Amateur Telescope Making*, edited by Albert Ingalls, served as my model, and it was driven by falling weights driving a piston into a cylinder of oil. This, like so many other concepts described in these volumes, seems quite old-fashioned and archaic now, but there remains so much of value in these books that they will probably always be an inspiration to new readers. It is also fascinating to see how far the sophistication and availability of astronomical equipment has become, compared to the lengths that our predecessors had to go to solve so many of the things we now take for granted. I can't help thinking that these pioneers enjoyed the finest hours in amateur astronomy with their new-found wonder and innovations.

If I were to build again, at this point a highly unlikely prospect, I would certainly be influenced by the design of the JMI NGT-18, which has been such a delight to use. The influence, and the main inspiration behind amateur telescope making, Russell Porter, remains as strong as ever in this design. It seems to address all of the engineering and optical issues that I wrestled with during my earlier days of telescope construction, though with a quality and sophistication that I could never have matched, or even approached. If you find yourself traveling down the homebuilt road, no matter how long, during your astronomical life, I can assure you of the positive effect it will have on your hobby. Maybe you will never escape its clutches, and for all the frustrations you may experience as you work to solve the problems unique to your creation, the influence this will have on your interest in astronomy will indeed be lifelong, regardless of whether you continue to build or ultimately succumb to one of the many commercially available solutions.

Bibliography

Burnham, Robert Jr., *Burnham's Celestial Handbook*, Volumes 1–3. Dover, New York, 1978.

Caidin, Martin and Barbree, Jay, *Destination Mars*. Penguin, New York, 1997.

Collins, W.J., Real Time Observation of Stellar and Galactic Sources using a Third Generation Image Intensified Optical System. Collins Electro Optics archive, Denver, Colorado, 1998.

Cook, Jeremy, Ed., *The Hatfield Photographic Lunar Atlas*. Springer-Verlag, 1968

Glass, Billy P., *Introduction to Planetary Geology*. Cambridge University Press, 1982.

Harrington, Philip S., *Star Ware*. John Wiley, 2002.

Ingalls, Albert G., Ed., *Amateur Telescope Making*, Volumes 1 and 2 (Volume 3 available). Scientific American, 1937.

Massey, S., Dobbins, T.A. and Douglass, E.J., *Video Astronomy*. Sky and Telescope Observer's Guides, 2000.

Menzel, D.H., *A Field Guide to the Stars and Planets*. Houghton-Mifflin, Boston, 1964.

Moore, Patrick, *Patrick Moore on Mars*. Cassell, London, 1998.

Paul, Henry E., *Telescopes for Skygazing*, second edition. Sky Publishing, Cambridge, Mass., 1966.

Peek, Bertrand M., *The Planet Jupiter*, revised edition. Faber and Faber, London, 1981.

Rudaux, Lucien and Vaucouleurs, G. De, *Larousse Encyclopedia of Astronomy*; revised edition. Paul Hamlyn, London, 1966.

Rükl, Antonin, *Atlas of the Moon*. Kalmbach Publishing, Wisconsin, 1996.

Sky and Telescope Magazine, Sky Publishing, 1978–2002.

Wilkins, H.P., *Moon Maps*. Faber and Faber, London, 1960.

Wilkins, H.P. and Moore, P., *The Moon*. Faber and Faber, London, 1961.

Index